The Commissioning and Routine Testing of Mammographic X-ray Systems

A protocol produced by a Working Party of the National Breast Screening
Quality Assurance Coordinating Group for Physics

A C Moore, D R Dance, D S Evans, C P Lawinski, E M Pitcher,
A Rust, K C Young

© Institute of Physics and Engineering in Medicine 2005
Fairmount House, 230 Tadcaster Road
York YO24 1ES
ISBN 1 903613 21 3

All rights reserved. No part of this publication may be reproduced, stored in a retrieval system or transmitted in any form or by any means, electronic, mechanical, photocopying, recording or otherwise, without the prior permission of the publisher.

Published by the Institute of Physics and Engineering in Medicine
Fairmount House, 230 Tadcaster Road, York YO24 1ES

Legal Notice

This report was prepared and published on behalf of the Institute of Physics and Engineering in Medicine (IPEM). Whilst every attempt is made to provide accurate and useful information, neither the IPEM, the members of IPEM or other persons contributing to the formation of the report make any warranty, express or implied, with regard to accuracy, omissions and usefulness of the information contained herein. Furthermore, the same parties do not assume any liability with respect to the use, or subsequent damages resulting from the use, of the information contained in this report.

Prepared and printed by:
York Publishing Services Ltd, 64 Hallfield Road, Layerthorpe, York YO31 7ZQ
Website: www.yps-publishing.co.uk

Contents

Acknowledgements (2005)		vii
Preface (2005)		viii
Preface to first edition of Report 59 (1989)		ix
Preface to second edition of Report 59 (1994)		x

1 Introduction — 1
  1.1 Mammography — 1
  1.2 Aims of this document — 1
  1.3 United Kingdom Breast Cancer Screening Programme — 2

2 Quality assurance — 4
  2.1 Introduction — 4
  2.2 Definition of terms — 4
  2.3 The quality system and the UK Breast Screening Programme — 5
    2.3.1 Management — 7
    2.3.2 Quality Assurance Manual — 7
    2.3.3 Quality Assurance Committee — 7
    2.3.4 Quality audit — 7
    2.3.5 Responsibilities — 7
  2.4 Limiting values, remedial and suspension levels — 8
    2.4.1 Limiting values — 8
    2.4.2 Remedial and suspension levels — 8
  2.5 Measurement uncertainties — 9
  2.6 Test equipment and calibration — 9

3 Physical principles of mammography — 10
  3.1 Introduction — 10
  3.2 Components of the mammographic system — 10
  3.3 Properties of the female breast — 11
  3.4 Interaction of photons with the breast — 12
  3.5 X-ray tube — 14
    3.5.1 X-ray spectrum — 14
    3.5.2 Focal spot size and imaging geometry — 20
  3.6 Breast compression — 22
  3.7 Anti-scatter grid — 22
  3.8 The mammographic screen–film receptor — 24

4 Digital imaging in mammography — 31
  4.1 Introduction — 31
  4.2 Photostimulable phosphors — 31

	4.3	Phosphor based direct digital systems	33
		4.3.1 Charge coupled devices	34
		4.3.2 Amorphous silicon/thin film transistor arrays	36
	4.4	Photoconductor based direct digital systems	37
	4.5	Comparison of digital and screen–film receptors	37
	4.6	Image display and digital processing	39
	4.7	Artefacts	40
5		The mammographic x-ray unit	42
	5.1	Introduction	42
	5.2	Electrical safety	43
	5.3	Mechanical safety and function	44
		5.3.1 Introduction	44
		5.3.2 Mechanical safety checks	44
		5.3.3 Marking and labelling	45
		5.3.4 Mechanical function checks	45
	5.4	Radiation safety	46
		5.4.1 Introduction	46
		5.4.2 Inspection	46
		5.4.3 Integral radiation protection screen	47
		5.4.4 X-ray room	48
	5.5	Additional checks for mobile equipment	48
		5.5.1 Before moving	48
		5.5.2 After moving	48
	5.6	Tests on the mammographic x-ray unit	49
		5.6.1 Introduction	49
		5.6.2 X-ray tube rating	49
		5.6.3 Alignment	49
		5.6.4 Leakage radiation	52
		5.6.5 Compression and breast thickness indication	53
		5.6.6 Dimensions of focal spot	54
		5.6.7 Tube voltage measurement	58
		5.6.8 Half value layer and filtration	59
		5.6.9 Tube output	60
		5.6.10 Anti-scatter grid	63
	5.7	Mammography system tests	64
		5.7.1 Introduction	64
		5.7.2 X-ray image uniformity	64
		5.7.3 Automatic exposure control system	67
		5.7.4 Automatic exposure control for small-field digital mammography systems	71
6		The testing of processors and screen–film systems	72
	6.1	Introduction	72
	6.2	Background to monitoring film processor performance	72
		6.2.1 Sensitometric parameters	74
		6.2.2 Applications of film processor sensitometry	75
		6.2.3 Causes of variation in sensitometric parameters	78

	6.3	The automatic processing unit	78
		6.3.1 Light sensitometry	79
		6.3.2 Temperature	79
		6.3.3 Transport speed	80
	6.4	Calibration of test equipment	80
		6.4.1 Calibration of densitometers	80
		6.4.2 Inter-comparison of sensitometers	80
	6.5	The cassette–screen–film system	81
		6.5.1 Screen–film contact	81
		6.5.2 Sensitivity matching of cassette–screen–film batches	82
	6.6	Darkroom and film storage conditions	83
		6.6.1 Subjective visual check	83
		6.6.2 Objective measurement	83
	6.7	Illuminators and viewing room conditions	85
		6.7.1 Subjective visual check	86
		6.7.2 Objective measurements	86
	6.8	Image artefacts	87
		6.8.1 System artefacts	87
		6.8.2 Film processing artefacts	87
7	Breast dose		89
	7.1	Introduction	89
	7.2	Dose specification	90
		7.2.1 Old standard breast model	90
		7.2.2 New standard breast model	91
	7.3	Estimation of mean glandular dose for the standard breast	92
	7.4	Estimation of mean glandular dose to real breasts	93
	7.5	Simulation of different thicknesses of breasts using Perspex	95
	7.6	Additional factors affecting mean glandular dose	96
		7.6.1 The consequences of multiple films or subsequent examinations	96
		7.6.2 Effect of magnification	96
		7.6.3 Stereotactic localisation	97
8	Image quality and test objects		98
	8.1	Introduction	98
	8.2	Clinical image quality	98
	8.3	Image quality assessment	99
	8.4	Image quality test objects	100
	8.5	Test object use	103
		8.5.1 Standard factors	103
		8.5.2 System resolution versus detector resolution	104
		8.5.3 Viewing conditions	104
	8.6	Image quality standards	105
		8.6.1 Limiting high contrast spatial resolution	105
		8.6.2 Threshold contrast	106
	8.7	Frequency of testing	108
	8.8	Description of test objects	109

9	Testing of stereotactic biopsy attachments		114
	9.1	Introduction	114
	9.2	Basic principles	114
	9.3	Testing of stereotactic localisation and biopsy devices	114
	9.4	Test objects	115
	9.5	Measurement procedure	118
	9.6	Problems	119
10	Testing of specimen cabinets used in mammography		121
	10.1	Introduction	121
	10.2	Calibration of equipment	121
	10.3	Safety checks	122
		10.3.1 Safety checks (HSE 2000)	122
		10.3.2 Radiation levels outside the cabinet	122
	10.4	Generator performance checks	122
		10.4.1 Tube voltage	123
		10.4.2 Tube output, linearity and consistency	123
		10.4.3 Exposure time	123
		10.4.4 Filtration	123
	10.5	Imaging performance checks	124
		10.5.1 Beam size and alignment	124
		10.5.2 Magnification factor	124
		10.5.3 Focal spot measurement	124
		10.5.4 Image quality	124
	10.6	Automatic exposure control tests (where fitted)	124

Appendix I: Suggested test frequencies — 125

Appendix II: Useful data — 127

Appendix III: Survey of performance measurements — 129

Appendix IV: Suppliers' addresses — 130

References — 135

ACKNOWLEDGEMENTS (2005)

The Institute is pleased to acknowledge the financial assistance given by the NHS Breast Screening Programme for the provision of the working party meetings. The Institute is also grateful to other societies, institutes, authors, manufacturers and publishers for allowing data and illustrations to be reproduced in this publication.

The authors are grateful to Ms A Burch for material relating to stereotactic test equipment, to Dr K Robson for material relating to the testing of viewing conditions and to Dr K Cranley for data on filtration.

PREFACE (2005)

This report revises and updates *The Commissioning and Routine Testing of Mammographic X-ray Systems* (IPEM Report 59) to take into account the latest developments in x-ray equipment technology including digital imaging, mammographic quality assurance and changes in legislation arising from the Ionising Radiations Regulations 1999 (IRR99), the Ionising Radiation (Medical Exposure) Regulations 2000 (IR(ME)R00) and the accompanying guidance.

The aims and applications of the protocol as set out in the original preface to Report 59 in 1989 remain unchanged for Report 89. The need to maintain uniformly high standards of equipment performance and of image quality for both screening and diagnosis remains of utmost importance.

Report 89 contains a new chapter on the theory and technology of digital imaging (Chapter 4). Chapter 5, relating to the testing of x-ray sets, has been substantially revised to take into account automatic beam optimisation software and multiple target/filter x-ray units and to emphasise the importance of looking at the x-ray imaging system as a whole. The testing of automatic processors (Chapter 6) has been rewritten to give the reader the background knowledge to some of the tests and to discuss system optimisation, recognising that the processor engineers or clinical staff are more likely to perform routinely most of the tests. The chapter on breast dosimetry (Chapter 7) has been re-written to incorporate changes to the standard breast model. The emphasis of the chapter on image quality and test objects (Chapter 8) has been changed to include a review of methods for assessment of image quality and to discuss test protocols and frequencies. The other chapters have also been revised to take into account new guidance and additional published work.

The report is intended to be a reference manual of tests that can be performed with the methodology and suggested frequencies that may be appropriate. It is not intended that every test in the report should be performed each time a unit is visited and guidance is given elsewhere regarding recommended minimum frequencies (IPEM, 1997).

PREFACE TO FIRST EDITION OF REPORT 59 (1989)

The production of this protocol has been stimulated by the introduction of a programme of breast cancer screening in the United Kingdom, following publication of the Forrest Report (DHSS, 1986). The Hospital Physicist's Association has already published a series of protocols (HPA 1980–1985) to help establish uniform methods of assessing diagnostic x-ray equipment, but these are not specific to mammographic x-ray equipment. Accordingly, a Working Party of the Diagnostic Radiology Topic group of the Institute of Physical Sciences in Medicine was established in October 1987 to produce a protocol for mammography. This document is the result.

The aim of the protocol is to provide methods for the commissioning and routine testing of mammographic x-ray systems which can be adopted nationally. It will be of use to medical physicists and medical physics technicians and also to x-ray engineers, radiographers, radiologists and other staff who have a responsibility for the testing of this equipment. However, it is recognised that at the present time there is only limited knowledge of quality assurance applied to mammography and that the procedures described in this document will need revision in the light of experience.

The protocol gives guidance on the quality system and includes checks and measurements on the x-ray machine, automatic processing unit and screen–film receptor and the assessment of the overall performance of the mammographic x-ray system. Where relevant the commissioning tests are dealt with first, followed by discussion of which ones are pertinent as routine tests. All the tests, together with suggested frequencies, are summarised in Appendix I. Throughout the text an attempt has been made to give a reasonable amount of practical detail and the style of a handbook has been adopted.

The Forrest Working Group stressed the importance of training of all the professions involved in providing the screening service. This protocol is an aid to the practical training of medical physicists, medical physics technicians, x-ray engineers, radiographers and other staff.

Although the need for guidance has arisen from the breast screening programme, it is hoped that this protocol will be used in the assessment of all mammographic x-ray units, whether they are used for screening or diagnosis, and so help to achieve uniformly high standards of performance, both in the National Health Service and in the private sector.

PREFACE TO THE SECOND EDITION OF REPORT 59 (1994)

When the first edition of this protocol was published in 1989 it was recognised that a second edition would be desirable within a few years which could incorporate experience gained from applying the protocol in the early years of the NHS BSP. The new edition has been prepared by a new Mammography Working Party of IPSM, assisted by comments from the UK Coordinating Group for Breast Screening Physics (the physics 'Big 18') consisting of one physicist from each English Region, each other UK country and the private sector.

The principal changes from the first edition include a completely new Chapter 3 by Dr D R Dance on the basic physics of mammography, substantial rewriting of parts of chapters relating to radiation dose and the testing of x-ray sets, film processors and associated equipment, and updating of the chapter describing test objects and phantoms. There are new chapters on stereotactic equipment and specimen cabinets. The opportunity to update references and organisational aspects has also been taken.

The aims and applications of the protocol as set out in the original 1989 Preface remain unchanged in 1994. The NHS BSP has been a major success but the need to maintain uniformly high standards of equipment performance and image quality both in screening and in diagnosis, is as necessary now as it was then.

CHAPTER 1

Introduction

1.1 Mammography

Mammography is the x-ray examination of the breast. Low energy x-rays are needed to provide adequate contrast in the image because the various tissues of the breast have similar attenuation coefficients (Johns and Yaffe, 1987). At the present time the majority of mammographic images are produced using a screen–film combination. The breast is exposed to x-rays generated at peak voltages in the range 25–32 kV from a molybdenum target tube using molybdenum, or other filter materials. Many units now provide the option of different target materials. Although screen–film combinations are still the most commonly used image receptors, other detector systems are currently in use or are still being developed (Yaffe and Rowlands, 1997). Photostimulable phosphor based computed radiography is used in some centres and small-field digital devices are used in stereotactic techniques and for spot imaging. A few full-field direct digital mammography systems are currently in use. Digital mammograms, produced either directly or by scanning conventional films, allow the use of computer aided analysis which shows promise as an aid to detection and diagnosis (Freer and Ulissey, 2001; Garvican and Field, 2001; Warren Burhenne et al., 2000).

Mammography is currently the examination of choice for routine breast screening and is used by the UK Breast Screening Programme. A review of mammographic imaging, physics and technique has been given by Säbel and Aichinger (1996). Ultrasound and MRI are used extensively for symptomatic disease and, in some situations, such as in younger women with dense breasts, for screening (Schnall, 2001; Warner et al., 2001).

1.2 Aims of this document

The technical demands on the mammographic imaging system are high to ensure optimal image quality for low radiation dose to the breast. The Ionising Radiation (Medical Exposure) Regulations 2000 (HMSO, 2000) put an obligation on employers to ensure that all diagnostic x-ray equipment is used in such a way as to reduce radiation dose to patients, or participants in health screening programmes, to the minimum consistent with achieving adequate diagnostic accuracy. In order to achieve this, the quality of each element of the imaging system needs to be assured.

The aim of this report, as of the two editions of Report 59 (IPSM, 1989, 1994), is to provide methods for the commissioning and routine testing of mammography x-ray systems which can be adopted throughout the United Kingdom. The report covers the performance testing carried out by medical physics services; additional, more frequent, routine tests will be carried out by the users of the equipment to verify stability. This revision has been produced due to considerable changes in both the technology and the standards required. An attempt has been made to provide enough scientific information for the reader to understand the subject and be able to interpret the results. Each test is described in sufficient detail for the purpose and method to be unambiguous and for this report to be used as a practical handbook.

The general principles of quality assurance and quality control are discussed in Chapter 2. The physical principles of mammography and the detection systems are described in Chapters 3 and 4. Chapters 5 and 6 give guidance on the tests and measurements required for the x-ray unit, automatic processing unit and image receptors. Chapters 7 and 8 give guidance on measurements of patient dose and image quality and allow the system performance to be assessed and related to national standards. Chapters 9 and 10 cover the testing of stereotactic units and specimen cabinets. Both commissioning and routine tests are described and the suggested test frequencies are summarised in Appendix I. Testing of ultrasound and MRI scanners is not covered in this report. The testing of this equipment is described in other documents (IPEM, 1998b; MDA, 1999a).

This report can be used as an aid to the training of medical physicists, medical technologists, x-ray engineers, radiographers and other staff. Its use should help to provide uniformly high standards of performance across the United Kingdom. The chapters on the physical principles and the protocols for the actual tests provide the basic theoretical knowledge required for medical physicists to act as medical physics experts in mammography as required by the Ionising Radiation (Medical Exposure) Regulations 2000 (HMSO, 2000).

1.3 United Kingdom Breast Cancer Screening Programme

In 1988 a national breast cancer screening programme was introduced in the United Kingdom within the National Health Service (NHS) following the publication of the Forrest Report (DHSS, 1986). The main conclusions of the Forrest Working Group were that women aged 50–64 years should be offered screening by mammography at three-year intervals and that high quality, single medio-lateral oblique view mammography should be used. More recent guidance in the NHS Cancer Plan for England has increased the age range to 50–70 years (DH, 2000).

The implementation of breast screening both in the UK and in other parts of the world stimulated the development of mammography and the search for improved techniques. The Department of Health and the Breast Screening Programme (NHSBSP) produced guidance notes on the requirements of mammographic equipment that have been regularly updated and expanded. The most recent is MDA Report 01/011 (DH 2001). The NHSBSP has produced performance standards for many aspects including those listed in Table 1.1 which are specific to mammographic imaging (NHSBSP, 2004). Note, the minimum standards quoted in the table are consistent with the remedial levels that are used in the rest of this report.

Table 1.1. NHSBSP quality standards relating to mammography

		Minimum standard	Target
To achieve optimum image quality	High contrast resolution (AP and LAT)	≥ 12 lp/mm	
	Minimum detectable contrast		
	5–6 mm detail	≤ 1.2%	≤ 0.8%
	0.5 mm detail	≤ 5%	≤ 3%
	0.25 mm detail	≤ 8%	≤ 5%
	Standard film density	1.5–1.9	
To limit radiation dose	Mean glandular dose per film to the standard breast at clinical settings	≤ 2.5 mGy	

Quality assurance is important to any radiographic examination (HMSO, 2000; WHO, 1982) and was seen by the Forrest Working Group as an essential element in the NHSBSP. It was considered vital, not just in relation to mammography, but to all aspects of the screening programme. Within the Ionising Radiations Regulations (HMSO, 2000), regulation 32(3) states that a suitable quality assurance programme should be in place for all equipment used for medical exposure. The accompanying approved code of practice states that when devising a suitable quality assurance programme for equipment the employer should give special attention to equipment used for medical exposure as part of a health screening programme.

Quality assurance is particularly important in mammography because the examination is so technically demanding. The NHSBSP explicitly required adequate arrangements for quality control within and between centres to maintain an acceptable standard of mammography. Quality assurance reference centres were set up in each breast screening region to control the quality assurance programme and, for each speciality, national coordinating groups were formed to deal with UK wide issues. The coordinating groups particularly relevant to mammography are physics, equipment, radiography and radiology. Quality assurance is discussed more fully in Chapter 2.

CHAPTER 2

Quality assurance

2.1 Introduction

The term 'quality assurance' has been applied increasingly to diagnostic radiology, but not always with a consistent meaning. Quality assurance has often been taken to mean routine performance testing of x-ray equipment, but its meaning is much broader and this was recognised in the Forrest Report (DHSS, 1986) and all additions to the NHS guidance on the establishment of a quality assurance system for the Breast Screening Programme (DH, 1989a; NHSBSP, 1999a, 1999b). An effective quality system (IEC, 1993b) will help to achieve and maintain:

(i) Radiological information of adequate quality for medical diagnostic purposes.

(ii) Minimum radiation dose to the patient and medical staff, compatible with adequate quality of the radiological information.

(iii) Maximum cost containment by minimising wastage of time and resources (e.g. reduction of rejected films).

(iv) Good professional practice.

The introduction of the Ionising Radiation (Medical Exposure) Regulations 2000 (HMSO, 2000) also puts an obligation on employers to ensure that all diagnostic x-ray equipment is used in such a way as to reduce radiation dose to patients, or participants in health screening programmes, to the minimum consistent with achieving adequate diagnostic accuracy.

2.2 Definition of terms

Quality assurance may be defined (BSI, 1991b) as all those planned and systematic actions necessary to provide adequate confidence that a product or service will satisfy given requirements for quality. Quality assurance has two components, namely: (i) *quality management*, which is that aspect of the overall management function that determines and implements quality policy; and (ii) *quality control*, which comprises the operational techniques and activities that are used to fulfil the requirements for quality.

In these definitions, *quality* refers to the total features and characteristics of a product or service that bears on its ability to satisfy stated or implicit needs, and the *quality policy* is the formal management statement as to the overall intentions and direction of an organisation with regard to quality. These terms apply equally to a radiology service, and the DH Guidelines (DH, 1989a) reflect this.

Quality management begins with the specification of an x-ray system and guidance has been given regarding mammographic screening equipment (MDA, 2001). The specification forms part of the procurement contract, together with general technical requirements (DH, 1989c). At installation, the equipment must undergo both a *critical examination* and *acceptance testing*. Regulation 31 of the Ionising Radiations Regulations 1999 (HMSO, 1999) puts the onus on the installer of the equipment to undertake a critical examination of the manner in which it was installed to ensure that safety features and warning devices operate correctly and that there is sufficient protection to persons from exposure to ionising radiation. The Health and Safety Executive (HSE) and the Institute of Physics and Engineering in Medicine (IPEM) have published guidance on carrying out a critical examination (HSE, 1998; IPEM, 1998a). Acceptance is the process of verifying that the contractor has supplied all of the equipment specified and has performed adequate tests to demonstrate that the specified requirements in the contract have been met (DH, 1989b; DHSS, 1985). The onus is on the contractor to undertake the tests. Acceptance may be a matter simply of completing a checklist. Mechanical and electrical safety tests are required in addition to radiation safety tests (NHSBSP, 2003b). Any significant discrepancies should be notified formally to the contractor who should be required to undertake corrective action.

Commissioning is the set of tests carried out by the customer's representative to ensure that the equipment is ready for clinical use and to establish baseline values against which the results of subsequent routine tests can be compared. Some of the commissioning tests take the form of checks and measurements; others are the process of optimising the performance of the imaging system. Commissioning tests are also known as status tests (IEC, 1988), although the latter may also refer to baseline tests on equipment that has been in use for some time.

Routine tests are those tests that are undertaken either regularly, or after maintenance or repairs, to detect whether any change in the performance of the equipment has occurred. After major work on the equipment, the relevant commissioning tests may have to be repeated to establish new baseline values.

The sequence of the tests is shown diagrammatically in Figure 2.1. Commissioning and routine tests together amount to the procedures and techniques of *quality control*. Guidance on quality control for the diagnostic x-ray imaging system in general is given by IPEM Report 77 (IPEM, 1997). The tests given here are equipment tests and therefore such matters as film reject analysis are not included. The scope and frequency of tests is outlined in Appendix I.

2.3 The quality system and the UK Breast Screening Programme

The establishment and operation of a quality system has been described (BS, 2000). Certain elements that have been adopted by the NHS Breast Screening Programme in England are detailed below (the programmes in Northern Ireland, Scotland and Wales have different arrangements but the same broad principles apply).

Figure 2.1. Flow chart of tests on x-ray equipment

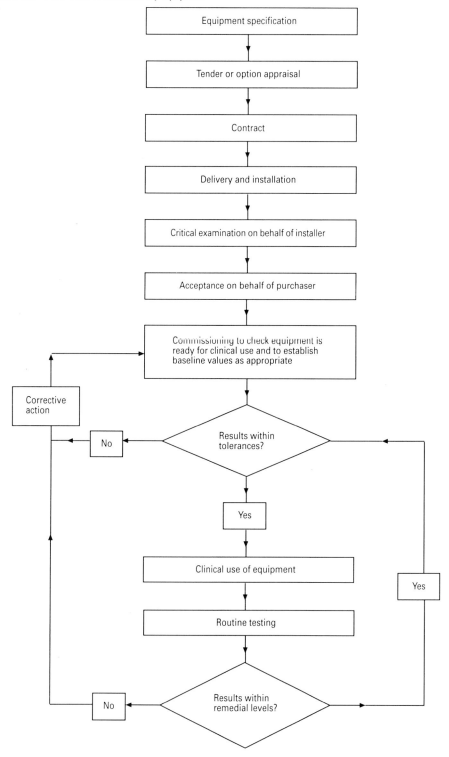

2.3.1 Management

Quality assurance is a management tool and requires a Quality Assurance (QA) Manager (DH, 1989a). The NHSBSP has given updated guidance on the organisation of quality assurance in the breast screening programme (NHSBSP, 2000). Quality assurance is led, in each region, by a regional director of quality assurance (QA Director) who is accountable to the regional director of public health. The regional QA Director appoints representatives from each of the main specialties involved in breast screening as QA coordinators. Regional quality assurance services are based on quality assurance reference centres (QARC). These are responsible for collecting and collating data, organising QA visits and supporting the QA Director and the professional coordinators. The QARC acts as the focus within a region for the development, monitoring and revision of the quality system. Other duties would be to prepare a quality assurance manual, to ensure that there is liaison between inter- and intra-regional units in the breast screening programme, and to ensure that the quality system is reviewed and audited at appropriate intervals. At unit level, a QA manager is appointed to perform similar functions locally.

2.3.2 Quality Assurance Manual

It is essential that the quality system is fully documented and that the manual reflects local conditions and organisation. Guidance on the contents of the QA manual was given by the Department of Health (DH, 1989a). The manual would be expected to specify the names and responsibilities of all persons involved in the regional quality assurance. It should also include details of acceptance, commissioning tests and routine tests including remedial and suspension levels, routine test frequencies and mechanisms of recording and reporting results. It may also specify the training required for all staff involved.

2.3.3 Quality Assurance Committee

This may be formed from all staff participating in the quality system, including the local and regional QA managers. It should provide overall guidance for the quality system and should ensure that any necessary changes are made and that the members of staff are kept informed.

2.3.4 Quality audit

This is a systematic and independent examination to determine whether the quality system is functioning effectively and whether the desired objectives are being achieved.

2.3.5 Responsibilities

Some of the responsibilities within the mammographic quality system are clear-cut; others are not so easy to define. A collaborative, multidisciplinary approach to quality assurance and performance testing is essential. However, the radiation safety aspects of new x-ray equipment are the responsibility of both the supplier and the employer whose staff use the equipment. The employer's Radiation Protection Adviser may also act on behalf of the supplier when carrying out the critical examination on new diagnostic x-ray equipment. Commissioning tests on x-ray equipment are normally undertaken by a medical physics service but radiographers will usually carry out tests on the film processors, films and cassettes. Those routine tests that have to be performed frequently are best done by the radiographic staff who use the mammographic x-ray system. Those routine tests that are necessarily less frequent, e.g. six monthly or longer, may be the province of the medical physics service, especially as they are more time-consuming and may require special test equipment and expertise. Feedback of the results to the operator is an essential element of a quality system.

2.4 Limiting values, remedial and suspension levels

2.4.1 Limiting values

At acceptance and commissioning, the performance of the equipment will be tested against the manufacturer's specifications and, if appropriate, the requirements given in MDA report 01/011 (MDA, 2001). The manufacturer's limiting values must be taken into account. These may be defined as the acceptable variation of the parameter being measured (CEC, 1989) and may also be referred to as tolerances (BSI, 1991a).

2.4.2 Remedial and suspension levels

A remedial level is that level of performance at which remedial action needs to be initiated. The action taken will depend on a detailed assessment of the equipment's performance and of the risk arising from its continued use. Following this assessment, agreement should be reached on any corrective action to be taken, the appropriate timescale for this action to be taken and any specific restrictions to be placed on the continued use of the equipment.

A suspension level is that level of performance at which it is recommended that the equipment should be removed from clinical use immediately until the performance is corrected. This may arise from either the results of routine testing or from a fault condition.

These definitions of remedial and suspension levels are taken from IPEM Report 77 (IPEM, 1997). The remedial and suspension levels have been based, where possible, on the effects on image quality and patient dose that the changes in a particular parameter may cause. In some cases there may be nationally decided guidance, e.g. the mean glandular dose to the standard breast should be less than 2.5 mGy, or they are based on changes from an established baseline level. These levels have been established from knowledge of normal variations and measurement uncertainties. Remedial and suspension levels are often represented by both positive and negative tolerances on the quantity in question, e.g. tube voltage. However, this is not necessarily the case; some quantities have an upper limit, e.g. breast dose, while others have a minimum threshold value, e.g. lead equivalence of a protective screen.

2.5 Measurement uncertainties

There are two categories of measurement uncertainty – systematic and random. Systematic uncertainties arise from physical effects which may influence or bias the result. Random uncertainties can be determined from an analysis of repeated measurements. Further details are given by Campion and co-workers (Campion *et al.*, 1980). The uncertainty associated with a measurement should be substantially less than the range represented by the limiting or remedial values and it is common to accept a measurement uncertainty of no more than one third of the range (CEC, 1989). However, the uncertainty associated with the estimation of some quantities, e.g. tube voltage, may approach the limiting values that ideally should apply and it may be some time before technical advances enable the uncertainty to be reduced.

Accuracy is the closeness of an observed quantity to the true value and precision is the closeness of agreement between the results obtained by applying a defined procedure several times under prescribed conditions (BSI, 1991a). In commissioning tests, good accuracy and precision are desirable to enable meaningful comparison with limiting values and to facilitate the intercomparison and compilation of data obtained from other systems or by different workers. In routine tests, accuracy is not so important, provided that the initial routine test is carried out as part of the commissioning; it is the precision of the measuring instrument and method that are important for these tests.

2.6 Test equipment and calibration

The required test equipment is listed as part of the test protocols in the relevant chapters and a list of current suppliers is given in Appendix IV. It is important that the equipment used should be appropriate to mammographic x-ray systems. For example, focal spot test tools such as the star test should allow accurate measurement of focal spot sizes of less than 0.5 mm, and dosimeters and kVp meters should be sensitive at low x-ray energies. Dosimeters and kVp meters should have a calibration traceable to a secondary standard laboratory. Calibrations should be carried out annually.

CHAPTER 3

Physical principles of mammography

3.1 Introduction

The quality of the mammographic image depends critically on the imaging equipment used and the way in which it is employed. In order to achieve and maintain a high image quality at a low breast dose, it is important to select mammographic equipment with an appropriate design and performance, and to use the correct operating parameters. This chapter provides an introduction to the physical factors that underpin the design, selection and use of the mammographic imaging system. Further discussion of the physical principles of mammography can be found, for example, in Säbel and Aichinger (1996) and Barnes and Frey (1991). Performance specifications for mammographic equipment are given in various Department of Health publications (DH 1989a, 1993, 1995, 2001a).

There are five physical parameters that must be considered when assessing the performance of the mammographic system: contrast, unsharpness, breast absorbed dose, noise, and dynamic range. Contrast is important because of the need to see small differences in soft tissue density. Because the breast is a small organ and there are no intervening soft tissue or bony structures, it is possible to use low energy x-rays to achieve this. Unsharpness is important because of the need to see small calcifications (Section 3.3). Dose must be kept low because of the risk of carcinogenesis associated with the examination (Beemsterboer *et al.*, 1998; Feig and Ehrlich, 1990; NCRP, 1986) and noise must be reduced because it affects the visibility of very subtle micro-calcifications. The dynamic range of the image receptor must be chosen so that it is sufficient to include the full range of breast tissues with adequate image quality. Each of these five parameters depends upon several components of the mammographic system. The relationship between them is complex and in general compromise is necessary.

3.2 Components of the mammographic system

The modern mammographic unit is designed to allow easy examination of the breast. The x-ray tube is mounted on a support arm together with the cassette holder and breast support, and the complete assembly may be elevated and rotated about a horizontal axis to achieve the desired radiographic projection.

The x-ray tube has a small focal spot and produces a low energy x-ray spectrum. The collimation of the radiation field and the position of the tube focus are arranged so that the field edge closest to the patient is vertical. This configuration ensures the visualisation of the maximum amount of breast tissue. The cathode–anode axis runs in the direction chest wall to nipple so that the heel effect provides most photons in the region where the breast is thickest and the photon transmission lowest.

The patient may be examined standing or sitting with her breast resting on a support table. Firm compression is applied using a plastic compression plate, to reduce the breast thickness and to hold it in the correct position so that the desired radiographic projection can be achieved. Both the compression plate and the breast support table should have a high x-ray transmission. The breast support plate forms the front face of a tunnel, which receives the image receptor in the form of a cassette containing a mammographic screen–film combination, or a digital receptor may be used. The table also incorporates an anti-scatter grid. The transmission of x-rays through the breast varies considerably, and automatic exposure control is essential. In many systems the x-ray spectrum is automatically selected based on breast thickness alone or on a combination of this and the transmission through the breast.

A magnification platform can be used to elevate the breast away from the image receptor, and in this case the x-ray tube is provided with a second, fine focus. The anti-scatter grid is removed during the magnification exposure as the air gap obtained when the breast is elevated provides sufficient scatter rejection. The equipment may also have accessories for stereotactic localisation and biopsy.

3.3   Properties of the female breast

The size and composition of the female breast vary widely. In infancy, the breast is composed primarily of adipose tissues, but at puberty fibroglandular tissue begins to develop, and this development continues until maturity. With further increase in age the fibroglandular tissue is gradually replaced by fat. Age is not, however, a strong indicator of tissue composition and wide variations of radiographic density occur in the breasts of women of similar age. Figure 3.1 shows typical values of breast glandularity for breast thickness in the range 2–11 cm and for women in the age range 50–64. The data (Dance et al., 2000a) show the decrease of glandularity with increasing breast thickness.

Figure 3.1. Estimates of average breast composition for different compressed breast thicknesses. Data are for women aged 50–64 attending the UK Breast Screening Programme (Dance et al., 2000a)

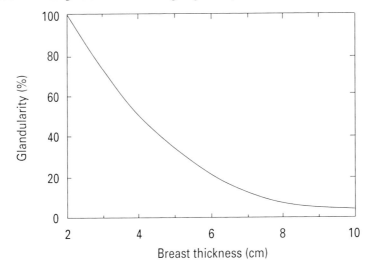

It is the glandular tissues within the breast (including acinar and ductal epithelium and associated stroma) which are believed to have a risk of radiation induced carcinogenesis. As a consequence, it is the mean dose to the glandular tissues within the breast which is normally used as a risk related dosimetric quantity in mammography (see Chapter 7 for a more detailed discussion of breast dosimetry).

The compressed breast can have a thickness of up to 11 cm with a median value of about 5.5 cm, depending upon population. Its area can be as small as 35 cm^2 or so large as to require the use of a cassette larger than the standard size (18×24 cm^2), or multiple exposures in order to image the whole organ. This range of compositions and sizes places high demands on the imaging system and the skill of the radiographer if good quality mammograms are to be obtained in all situations.

Table 3.1 gives the atomic compositions of the adipose and glandular tissues within the breast. Knowledge of the composition of these tissues and the relative amounts of each present in the breast is very important for any study of the physical properties of the mammographic system.

Table 3.1. Composition of the adipose and glandular tissues within the breast. Data are taken from Hammerstein et al. (1979)

	Percentage composition by weight			Density (g cm^{-3})
	H	C	O	
Fat	11.2	49.1–69.1	18.9–35.7	0.93
Gland	10.2	10.8–30.5	55.2–75.9	1.04

When early breast cancer is detected, there is often no mass visible on the mammogram and the only visible abnormality is a cluster of micro-calcifications. Calcifications are specks of calcium hydroxyapatite or calcium phosphate, which are associated with breast cancer in up to 50% of cases (Millis et al., 1976). They vary in size from less than 50 μm to several mm. It is therefore important that the mammographic imaging system has a high resolution. In practice calcifications of size 100 μm upwards may be visible on the mammogram.

3.4 Interaction of photons with the breast

The compositions and densities of the adipose and fibro-glandular tissues within the breast are quite different, and this is reflected in their absorption and scattering of x-ray photons. Figure 3.2 shows the variation of the linear attenuation coefficient with photon energy for adipose and fibrous tissues and for infiltrating duct carcinoma (Johns and Yaffe, 1987). It demonstrates that the highest contrast is obtained by imaging using low energy photons. It also shows that there is only a small difference between the photon interaction properties of carcinoma and fibrous tissue. In fact, the lines plotted represent averages from a series of tissue samples, and, above 28 keV, there is some overlap between the results obtained from individual samples of cancerous and fibrous tissues. The consequence is that breast cancers may appear on the mammogram as tissues with a radiographic density not very different from that obtained for normal structures, and may be identified only as a disturbance to the normal breast architecture.

Figure 3.2. Dependence of the linear attenuation coefficients for three breast tissues on photon energy: A, infiltrating duct carcinoma; B, fibrous tissue; C, fatty tissue. Data taken from Johns and Yaffe (1987) with kind permission of IOP Publishing Ltd

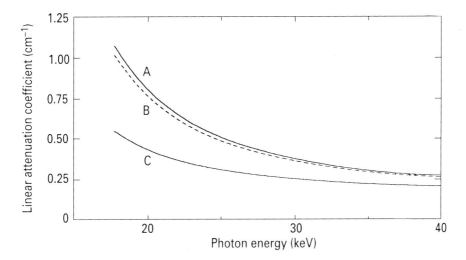

In the mammographic energy range, the principal interactions of x-ray photons with tissue are the photoelectric effect and photon scattering. Figure 3.3 shows the contributions of these interactions to the linear attenuation coefficient for photon energies in the range 15–40 keV. The photoelectric effect is the dominant interaction below 22 keV. It is this interaction that is responsible for most of the energy which is imparted to the breast by the incident photons. For scattering processes, the photon energy is sufficiently small that atomic and inter-atomic binding of electrons cannot be neglected (Dance, 1980) and the scattering is treated as a sum of coherent and incoherent processes. In coherent scattering, all the electrons scatter in phase and there is no energy transfer to the breast. This process is largest in or close to the forward direction, where it is enhanced by a factor of Z (atomic number) and modified by the effects of inter-atomic binding (Evans et al., 1991; Kidane et al., 1999). For incoherent or Compton scattering, there is an energy transfer, but in the mammographic energy range this remains small. (Backscattering of a 30 keV photon provides an energy transfer of only 3.2 keV.) In spite of the small energy transfer to the breast, photon scattering remains of considerable importance in mammography. This is because of its contribution to the primary contrast in the image (as one of the interaction processes contributing to the total interaction probability) and because of the contrast degrading effect of scattered photons recorded by the image receptor.

Figure 3.3. Dependence of the linear attenuation coefficients for breast tissue (equal parts by weight adipose and glandular tissue) on photon energy. The upper and lower curves give the attenuation coefficients corresponding to all interactions (i.e. the sum of photoelectric and scatter processes) and just scatter processes respectively

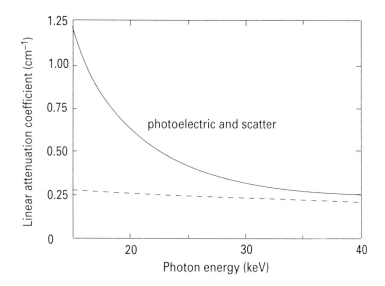

3.5 X-ray tube

3.5.1 X-ray spectrum

The choice of the x-ray spectrum for mammographic imaging is a compromise between the requirements of high image quality and low breast dose. Image quality improves as the photon energy is reduced (both contrast and the signal-to-noise-ratio, for fixed energy absorbed per unit area of the receptor, increase) whereas the dose becomes larger. This is illustrated in Figures 3.4, 3.5 and 3.6, which show how these quantities vary with photon energy. The contrast in Figure 3.4 has been calculated using a simplified model (see figure caption), and is given for a 100 µm calcification and for 1 mm of glandular tissue. Both contrast curves show a rapid decrease with increasing photon energy: there is a factor of 6 decrease between energies of 15 and 30 keV.

Figure 3.5 illustrates the dependence of the mean glandular breast dose on photon energy and breast thickness. The curves have been calculated for a typical mammographic screen–film combination and for breasts of thicknesses 2 and 8 cm with compositions in accordance with Figure 3.1. They show the severe dose penalty incurred when a low photon energy is used to image a large breast.

Figure 3.4. Dependence of contrast on photon energy. The contrast has been calculated using a very simple model which neglects scatter, unsharpness and receptor gain, using the general approach and the definition of contrast adopted in Dance (1988). The upper curve is for a 100 μm calcification viewed against a background of 'average' breast tissue and the lower curve is for 1 mm glandular tissue viewed against a background of adipose tissue

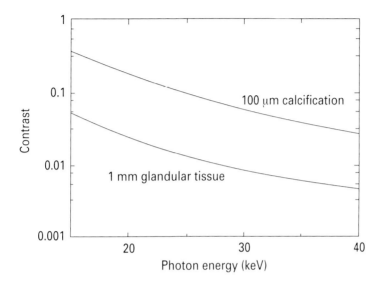

Figure 3.5. Dependence of mean glandular breast dose on photon energy and breast thickness. Upper curve: compressed breast of thickness 8 cm and glandularity 7%. Lower curve: compressed breast of thickness 2 cm and glandularity 100%. Results based on Monte Carlo calculation (Dance, unpublished) for a typical mammographic screen–film receptor. The data have been normalised to the value for the 2 cm breast at 20 keV

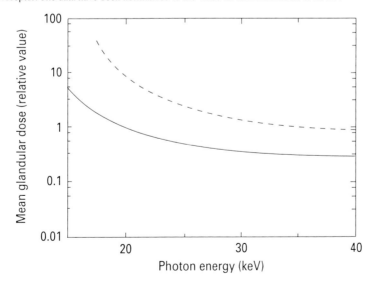

16   *The Commissioning and Routine Testing of Mammographic X-ray Systems*

The signal-to-noise ratio (SNR) for imaging a 100 μm calcification is shown in Figure 3.6. This has been calculated for a mammographic screen–film combination at an optical density of 1.0 using a very simple model. It is assumed that the only noise contribution arises from quantum mottle. The effects of image unsharpness, film granularity and screen-structure mottle have been ignored. In spite of these approximations, it is clear that there is a rapid decrease in SNR with increasing photon energy and that there is a threshold energy above which the calcification will not be visualised.

Figure 3.6   Dependence of signal-to-noise ratio on photon energy when the energy absorbed per unit area of the receptor is fixed. Signal calculated for imaging a 100 μm calcification, neglecting scatter, unsharpness and receptor gain. Noise only includes contribution from quantum mottle. Mammographic screen–film receptor

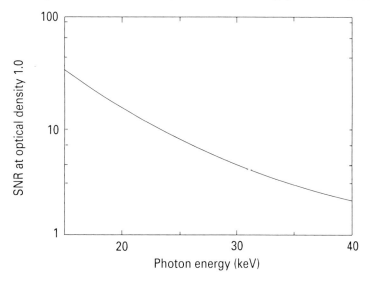

Various authors have discussed how to combine the above energy variations to predict the optimal energy or energy spectrum for mammography (e.g. Dance and Day, 1981; Dance *et al.*, 2000b; Desponds *et al.*, 1991; Fahrig and Yaffe, 1994; Gingold *et al.*, 1995; Jennings and Fewell, 1979). It is often assumed that mammography is a noise-limited imaging process. The approach then adopted is to set the signal-to-noise ratio for imaging a particular object at the required value for adequate visualisation and to find the photon energy or spectrum that achieves this value at the minimum breast dose. Figure 3.7 shows the results of such a calculation for imaging a 100 μm calcification. Four curves are shown corresponding to compressed breast thicknesses of 2, 4, 6 and 8 cm. Each curve passes through a minimum value, which gives the optimum photon energy, and the position of the minimum varies with photon energy. Optimum energy bands can be derived from this figure for imaging the above breast thicknesses, and these are given in Table 3.2.

Table 3.2. Optimal energy bands for mammography based on dose for fixed SNR

Breast thickness (cm)	2	4	6	8
Energy band (keV)	14–18	17–21	19–23	20–26

Figure 3.7. Calculations of the mean glandular dose for imaging a 100 μm calcification at a fixed signal-to-noise ratio of 5 with a typical mammographic screen–film receptor. Results are shown for breast thicknesses of 2, 4, 6 and 8 cm. In each case the breast has glandularity 50%. Data calculated using the simplified methodology of Dance et al. (2000a) which neglects image unsharpness and assumes that quantum mottle is the only noise source

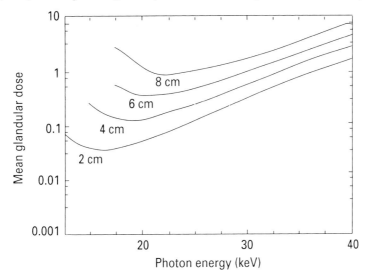

The x-ray spectrum that is conventionally used for screen–film mammography is shown in Figure 3.8 (A and B) before and after passage through 5 cm tissue. It is produced by an x-ray tube with a molybdenum target and is filtered by a molybdenum foil 30 μm thick. Its principal features are the molybdenum characteristic x-ray lines at 17.4 and 19.6 keV and the sharp cut-off above 20.0 keV, which is the position of the K-absorption edge of molybdenum. Peak voltages of 25–30 kV are normally used, with the choice depending upon breast thickness and local preference. Molybdenum anode tubes have a beryllium exit window to avoid undue hardening of the beam and the 30 μm molybdenum filter provides the required minimum beam filtration. Reference to Table 3.2 shows that this spectrum is well suited to imaging all but the largest breasts, for which a spectrum with a slightly higher energy would offer some dose advantage. Such spectra can be obtained by replacing the molybdenum filter with a filter of a slightly higher atomic number such as rhodium (K-edge 23.2 keV). It is also possible to use an x-ray tube with a rhodium (the rhodium characteristic x-ray lines are at 20.2 and 22.8 keV) or tungsten anode. Examples of these alternative spectra before and after 5 cm of tissue are illustrated in Figure 3.8 (C–H).

Figure 3.8. Mammographic x-ray spectra at 28 kV from various target filter combinations: (A, B) molybdenum target 30 μm molybdenum filter; (C, D) molybdenum target 25 μm rhodium filter; (E, F) rhodium target 25 μm rhodium filter; (G, H) tungsten target 50 μm rhodium filter. Spectra A, C, E and G are spectra incident on the breast, and spectra B, D, F and H are primary spectra after passage through 5 cm of tissue. Spectra were calculated using data of Cranley *et al.* (1997). Each spectrum is normalised to the same maximum value

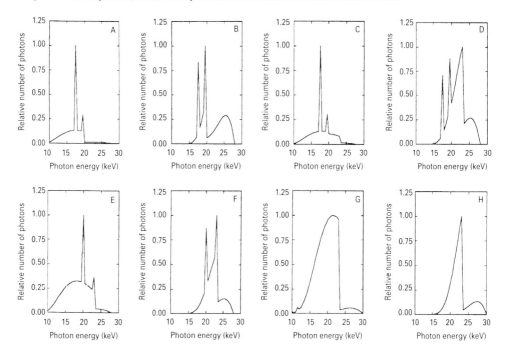

The above discussion of optimal spectra rests on the assumption that mammographic imaging is noise limited. For screen–film imaging, where the exposure level is determined by the need to obtain adequate optical density on the film, and it is not possible to manipulate the image after it has been captured, the contrast in the image is also of considerable importance. Figures 3.9 and 3.10 show model calculations of contrast and mean glandular dose for imaging breasts 4 cm thick (glandularity 50%) and 8 cm thick (glandularity 10%) with spectra from molybdenum/molybdenum (Mo/Mo), molybdenum/rhodium (Mo/Rh), rhodium/rhodium (Rh/Rh) and tungsten/rhodium (W/Rh) target/filter combinations (Dance *et al.*, 2000b). Each curve in these figures corresponds to a particular target/filter combination and to tube voltages in the range 25–32 kV. It will be seen from the results for the 4 cm thick breast that none of the spectra can match the contrast achieved using a Mo/Mo spectrum at 29 kV or below. Since at this breast thickness a tube voltage of 28 kV or less would normally be used, the alternative spectra only offer a dose saving if a loss in contrast compared with the conventional technique is acceptable. As the breast thickness changes, so does the comparison between the different spectra. Figure 3.10 shows that for an 8 cm thick breast, the contrast achievable using at Mo/Mo spectrum at 28 kV can be matched using a Mo/Rh spectrum at 25 kV and with some dose saving. In practice, the choice of spectrum will depend upon local preference and will be influenced by the shape of the film characteristic curve. If a screen–film combination with a relatively high gamma is used, it may be acceptable or even desirable to accept some reduction in contrast (Young *et al.*, 2001).

Figure 3.9. Relationship between mean glandular dose and the contrast for imaging a 200 μm calcification for a model breast 4 cm thick and 50% glandularity. Curves are shown for four target/filter combinations. The points plotted for each combination correspond to tube voltages in the range 25–32 kV. Data based on Dance *et al.* (2000b) and reproduced with permission of the *British Journal of Radiology*

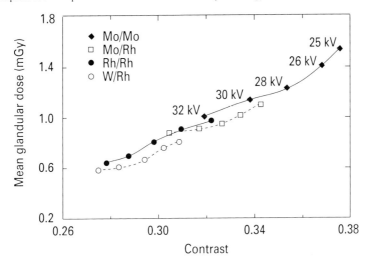

Figure 3.10. Relationship between mean glandular dose and the contrast for imaging a 200 μm calcification for a model breast 8 cm thick and 10% glandularity. Curves are shown for four target/filter combinations. The points plotted for each combination correspond to tube voltages in the range 25–32 kV. Data based on Dance *et al.* (2000b) and reproduced with permission of the *British Journal of Radiology*

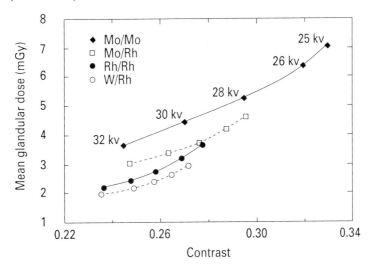

Image receptors for digital mammography have a wide dynamic range, and the requirement in screen–film imaging to control the optical density no longer applies. Optimal spectra can be chosen on the basis of the dose required to achieve a given signal-to-noise ratio. This is provided that any loss of image contrast associated with using a harder spectrum can be recovered by suitable image processing. Figures 3.11 and 3.12

show simple model calculations of the mean glandular dose for the task of imaging a 200 μm calcification at an SNR of 5 (Dance et al., 2000b). Each figure gives the results of calculations for Mo/Mo, Mo/Rh, Rh/Rh and W/Rh spectra as a function of tube voltage. Figure 3.11 is for a 2 cm compressed breast of 50% glandularity, and shows that for this case, the Mo/Mo spectrum is a good choice. As the breast thickness increases, however, other target filter combinations can offer a dose saving at fixed SNR. Figure 3.12 shows this for an 8 cm breast of 10% glandularity. In this case the W/Rh spectrum offers the lowest dose at fixed SNR and the dose for the Mo/Mo spectrum is actually the highest of the four spectra considered.

Figure 3.11. Relationship between the mean glandular dose and tube voltage for imaging a 200 μm calcification at an SNR of 5 for a model breast 2 cm thick and 50% glandularity Curves are shown for four target/filter combinations. Data based on Dance et al. (2000b) and reproduced by permission of the *British Journal of Radiology*

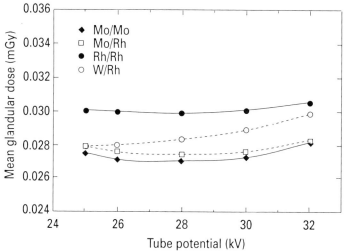

3.5.2 Focal spot size and imaging geometry

There are four factors that determine the overall unsharpness in the mammographic image: the focal spot size, the imaging geometry, the receptor performance and the patient movement. If the breast is well compressed, the exposure time is short and there is no mechanical vibration from the movement of the grid, movement unsharpness should be small and it is neglected in the present analysis; The first two factors influence the geometric unsharpness, and the overall unsharpness can be studied by combining this quantity with the receptor unsharpness. If it is assumed that these two contributions can be combined in quadrature, then the overall unsharpness U (expressed at unit magnification), is given by:

$$U = \frac{1}{M}\sqrt{(M-1)^2 f^2 + F^2} \qquad (3.1)$$

where M is the image magnification, f the focal spot size and F the receptor unsharpness. The geometric unsharpness, which is given by the first term on the right hand side, is zero for magnification unity.

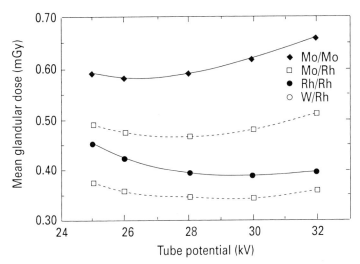

Figure 3.12. Relationship between the mean glandular dose and tube voltage for imaging a 200 μm calcification at an SNR of 5 for a model breast 8 cm thick and 10% glandularity Curves are shown for four target/filter combinations. Data based on Dance et al. (2000b) and reproduced by permission of the *British Journal of Radiology*

Figure 3.13 shows Equation 3.1 evaluated as a function of focal spot size and magnification for receptor unsharpness of 100 μm. It is evident that magnification will increase the overall unsharpness unless a sufficiently small focal spot size is used, and magnification mammography can only be recommended when a fine focus is available. The figure shows that there is very little improvement in unsharpness if a small focal spot is used and the magnification increases from 1.5 to 2.0. When selecting the most appropriate magnification, however, consideration must also be given to image noise, the increased patient dose if the same receptor is used, and the effect of the air gap on the amount of scattered radiation reaching the receptor.

For conventional (non-magnification) mammography, the figure demonstrates that a quite small focal spot is still required unless a magnification close to unity can be achieved. This implies that a large focus receptor distance should be used with the breast positioned as close as possible to the image receptor. In practice, for the top surface of a breast at (for example) 6 cm from the receptor and a focus receptor distance of 65 cm, the image magnification is 1.10 and the focal spot size of 0.5 mm (measured), the maximum recommended in DH (2001a), can give very good performance. For magnification, DH (2001a) recommends a measured focal spot size of ≤0.15 mm. An important factor, which limits the size of the focal spot, is the requirement of a high tube output, so that the exposure time can be as short as possible to limit the effects of patient movement and reciprocity law failure. DH (2001a) recommends tube currents of at least 100 mA and 30 mA respectively on broad and fine focus at 28 kV.

Figure 3.13. Dependence of overall unsharpness on image magnification and focal spot size. The receptor unsharpness is taken as 100 μm. The unsharpness is given for focal spot sizes of 0, 100, 500 and 1000 μm

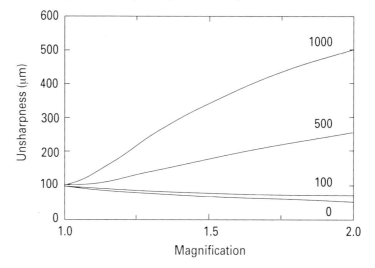

3.6 Breast compression

Breast compression is essential in mammography because it reduces absorbed dose and image unsharpness, and it increases image contrast. Dose is reduced because of the increased penetration of the x-ray photons through thinner tissue sections (Figure 3.5). Image unsharpness is reduced for two reasons. Firstly, most of the structures in the breast will be brought closer to the image receptor, thereby reducing geometric unsharpness (Section 3.5.2), and secondly, the breast will be held firmly in place and the exposure time will be reduced, both of which will decrease movement unsharpness. Contrast is also improved for two reasons. Firstly, the quantity of scattered radiation leaving the bottom of the breast is reduced; and secondly, the x-ray spectrum reaching the image receptor is softer and the primary contrast higher. A further important aspect of breast compression is that the breast architecture is imaged over a larger projected area. This reduces the overlap of structures in the image and makes the breast architecture easier to interpret. Compression may also help to demonstrate architectural distortion in the location of a tumour. Finally, because the tissue path length will be similar for much of the image and the range of path lengths less, it is easier to fit the image within the latitude of the image receptor. This may also have the advantage of allowing the use of an image receptor with a higher contrast gain.

3.7 Anti-scatter grid

The contrast in the mammographic image is degraded by the scattered radiation recorded by the image receptor. The amount of this degradation varies with photon energy, and with breast size and composition. It can be quantified using the contrast degradation factor, *CDF*, which is the ratio of the image contrasts obtained with and without scattered radiation. The *CDF* depends on the ratio of the energies imparted per unit area of the receptor by scattered and primary radiation (*S/P*) and is given by:

$$CDF = 1/(1 + S/P) \qquad (3.2)$$

Figure 3.14 shows the dependence of *S/P* on breast thickness at 28 kV and a compressed breast area of 100 cm^2 (Dance *et al.*, 1992). The contrast degradation factor (Equation 3.2) shows some dependence on breast composition and is 0.62, 0.65 and 0.67 for 5 cm thick breasts composed of pure adipose tissues, equal parts by weight adipose and glandular tissues, and pure glandular tissues respectively. The variation of the *CDF* with tube potential is small and amounts to a difference of less than 2% between 25 and 30 kV. The effect of breast area on *S/P* is discussed by Dance and Day (1984). For monoenergetic incident photons at 25 keV and a 6 cm thick breast, *S/P* changes by 7% as the breast area varies between 35 cm^2 and 270 cm^2.

Figure 3.14. Dependence of the scatter-to-primary ratio on breast thickness. 28 kV Mo/Mo spectrum. Results given for examinations without a grid and with a typical focused linear mammographic grid

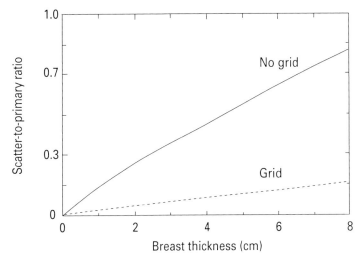

The use of grids to reduce the amount of scatter recorded by the image receptor is well established. Two types of grid are presently used which are specially designed for mammography. Most equipment uses a moving linear focused grid with a line density of around 30 lines/cm, a grid ratio of 4–5 (the grid ratio is defined as the ratio of height of the lead strips to the spacing between them) and low atomic number material for the grid interspaces and covers. A detailed study of the performance of such a grid in various imaging situations is given by Dance *et al.* (1992). Figure 3.14 shows the scatter-to-primary ratio these authors obtained with a typical linear focused grid and Figure 3.15 shows the corresponding contrast improvement factors (*CIF*) and grid factors (*GF*). The *CIF* is the ratio of the contrasts obtained with and without the grid, and the *GF* is the corresponding ratio for the exposures. The *CIF* takes due account of the beam hardening (and associated small loss of contrast) by the grid (Alm Carlsson *et al.*, 1986). For a large breast (8 cm thick) the particular linear focused grid studied gave a 54% increase in contrast at a grid factor of 2.1. For a small breast (2 cm thick) the improvement was less, amounting to 17% at a grid factor of 1.7. For a 5 cm thick breast imaged with the focused linear grid, the image contrast and breast dose decrease by 17% and 29% respectively as the tube voltage is increased from 25 to 30 kV.

Figure 3.15. Dependence of the contrast improvement factor (*CIF*) and grid factor (*GF*) on breast thickness for a typical focused mammographic grid. 28 kV Mo/Mo spectrum

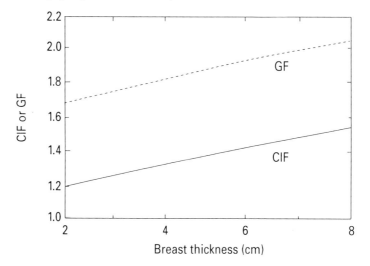

An alternative to the focused linear grid is the High Transmission Cellular™ (HTC) grid introduced by Lorad. Linear grids are only effective in reducing scatter that intersects the lead strips. They have a high transmission for photons travelling approximately parallel to the strips. The HTC grid has a focused cellular structure which can in principle offer a low transmission for almost all scattered photons (apart from those close to the forward direction). It has sets of copper strips running in two perpendicular directions and air interspaces between the strips. When such a structure is moved during the exposure, it will in general produce image artefacts corresponding to the positions where the two sets of strips overlap. However, the manufacturers have developed a precise movement control for the grid that avoids any visible artefacts. Rezentes *et al.* (1999) have studied the performance of the HTC grid using test phantoms. For tube voltages of 25 and 30 kV, they found that the CIF was 5–10% better than that for a conventional linear focused grid.

3.8 The mammographic screen–film receptor

The speed of a mammographic screen–film system depends upon many factors, some of which also influence noise, contrast and resolution. Some early mammographic screens (Ostrum *et al.*, 1973) used a calcium tungstate phosphor, but current screens use rare earth phosphors. Figure 3.16 and Table 3.3 compare the properties of a selection of phosphors. The figure gives the energy absorption efficiency for these phosphors calculated for the same screen thickness and packing density, so that comparison can be made at the same resolution. Gadolinium oxysulphide and calcium tungstate have the best efficiency curves. The table shows the efficiency with which the energy absorbed from the incident x-ray is converted to light fluorescent photons, and the K-edges of the phosphors. The rare earth phosphors have the highest conversion efficiencies, and on the basis of the figure and table, gadolinium oxysulphide is the phosphor of choice. It should be noted, however, that in practice it may not be possible to use all of the light output from the screen as dyes may be added to improve the resolution, thereby decreasing the quantum noise in the image at a given optical density.

Figure 3.16. Dependence of energy absorption efficiencies of various phosphors on photon energy. Curves are shown for $CaWO_4$ and Gd_2O_2S (superimposed) (solid curve), La_2O_2S (dashed curve) and Y_2O_2S (dotted curve). Each curve is for a screen 100 μm thick with 50% packing density. Data taken from Dance and Davis (1983) with kind permission of Chapman and Hall

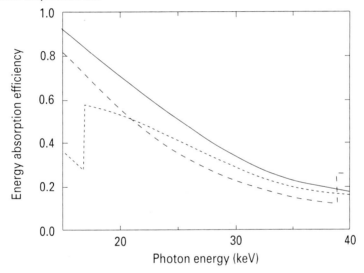

Table 3.3. Properties of screen phosphors. Light efficiencies taken from Stevels (1975)

Phosphor	K-edge (keV)	Light emission efficiency (%)
$CaWO_4$	69.5	3.5
Gd_2O_2S	50.2	15
La_2O_2S	38.9	12
Y_2O_2S	17.0	18

Figure 3.17 shows the energy absorption efficiency of a typical mammographic screen, using a gadolinium oxysulphide phosphor. The energy absorption efficiency is moderately high at the low energies used in mammography, but has been restricted because of the need to achieve high resolution. The efficiency for absorption of scattered photons is higher than that for primary photons because the former have a greater path length in the receptor and a slightly lower photon energy. An inefficient receptor enhances the contribution of scattered photons to the image.

Figure 3.17. Dependence of energy absorption efficiency of a typical mammographic screen on photon energy. Efficiencies are shown for both primary (solid curve) and secondary (dashed curve) photons. Data taken from Dance and Day (1984) by kind permission of IOP Publishing Ltd

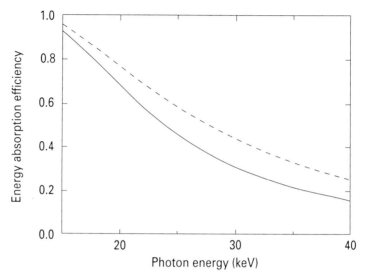

The fluorescent light photons produced by the screen are emitted isotropically and spread out laterally as they pass from their point of emission to the film emulsion. The spread increases with the distance travelled, so that resolution can be improved by reducing this distance. This is best achieved using a single screen and a single emulsion film with the screen positioned after the film and in close contact with the emulsion. This brings the point of x-ray absorption/light emission as close as possible to the emulsion. The use of a single emulsion also prevents the crossover effect associated with double screen/double emulsion receptors. Figure 3.18 shows the modulation transfer function for a typical mammographic screen–film system.

Figure 3.18. Modulation transfer function for a Kodak MinR 2000 mammographic screen–film combination. Data taken from Bunch (1999) and reproduced by kind permission of SPIE

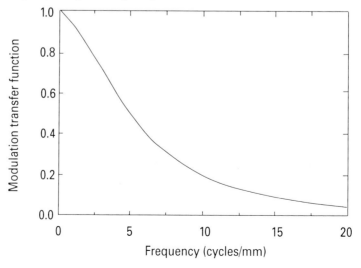

Mammographic screen–films must be matched to the emission spectrum of the phosphor and are designed to have a high contrast and narrow latitude. They give an optical density in excess of 3.0 in regions close to the edge of the breast; so that the mammogram must be viewed in a darkened room using a high intensity light box with all glare masked off. Figure 3.19 (A) shows characteristic curves for two mammographic screen–films and Figure 3.19 (B) shows the gradient (film gamma) of these characteristic curves as a function of optical density. With such steep characteristics, it can be difficult to fit the image within the steepest part of the curve, so that the entire image is recorded in a density region where the film gamma is high. Accurate automatic exposure control and the monitoring of film processing are therefore essential. Too low or too high an exposure will result in parts of the breast being imaged in the toe or shoulder regions of the curve, thus leading to inadequate contrast. A practical difficulty for the AEC system is the compensation of reciprocity law failure, which can be significant for mammographic films. For example, the results of de Almeida et al. (1999) for three different mammographic screen–film combinations show increases of about 20–25% in the exposure required for a gross optical density of 1.5 as the exposure time increases from 0.5 s to 5 s.

Figure 3.19. Film characteristic curve (A) and film gamma (B) for Fuji UM-MA (solid curve) and Kodak MinR2000 (dashed curve) mammographic screen–film combinations. Data taken from Meeson et al. (2001) and reproduced with kind permission of the British Institute of Radiology

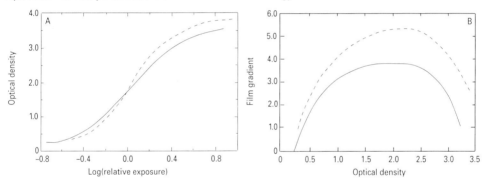

The shape of the film characteristic curve and the film speed depend upon the film processing time, processing chemicals and developer temperature. In the past, the use of an extended processing time was popular as this could reduce dose whilst maintaining or improving contrast (Skubic et al., 1990; Tabar and Haus, 1989). Kimme-Smith (1991) provides a good summary and explanation of these options. More recently screen–film combinations have been developed that can achieve a high contrast and a low dose with a shorter processing cycle, so that extended processing is not necessary. Such screen–film combinations tend to work best in modern processors that can ensure adequate fixing, washing and drying.

The noise in the image produced by a screen–film receptor arises from five principal sources:

(i) Fluctuations in the number of x-ray photons interacting per unit area of the screen (quantum mottle).

(ii) Fluctuations in the energy absorbed per interacting photon.

(iii) Fluctuations in the number of light fluorescent photons emitted per unit x-ray energy absorbed that are successful in reaching the emulsion.

(iv) Fluctuations in the number of silver halide grains available per unit area of the emulsion and in the size and sensitivity of individual grains (film granularity).

(v) Fluctuations in the screen absorption associated with inhomogeneities in the phosphor coating (structure mottle).

The most important of these are usually the quantum mottle and the film granularity. Noise contributions (ii) and (iii) can be regarded as making modifications to the magnitude of the quantum mottle.

Figure 3.20 shows the noise power spectrum for the Kodak MinR/Ortho M screen–film combination measured by Nishikawa and Yaffe (1985) at an optical density of 1.0. The quantum mottle is the largest contribution except at the higher frequencies where it is not resolved by the system. At high or low optical densities, however, film granularity becomes much more significant because the quantum mottle is proportional to the gradient of the film characteristic curve, which is much reduced at these densities (Barnes and Chakraborty, 1982).

Figure 3.20. Noise power spectrum for the Kodak MinR-OrthoM screen–film combination at an optical density of 1.0. Data taken from Nishikawa and Yaffe (1985) with kind permission of the American Association of Physicists in Medicine. The solid curve represents quantum noise power and the dashed curve film noise power

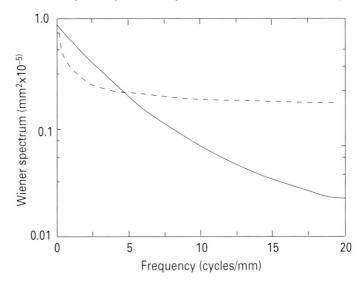

The detective quantum efficiency (DQE) is a very useful measure of the performance of the image receptor, which takes into account the combined effects of all noise sources (ICRU, 1995). It is defined by the relationship:

$$DQE = \left(\frac{SNR_{out}}{SNR_{in}}\right)^2 \quad (3.3)$$

where SNR_{in} and SNR_{out} are the signal-to-noise ratios of the pattern of x-ray photons incident on the image receptor and of the image respectively. It thus gives information about the transfer of SNR by the receptor. For our purposes it is helpful to express the DQE as

$$DQE = \varepsilon f \quad (3.4)$$

where ε is the probability of an incident photon interacting with the image receptor and f is a factor less than or equal to unity. If all the interacting photons contribute the same signal to the image, then $f=1$ and the DQE is equal to the photon interaction probability. In practice this is not the case (because of the additional noise sources identified above); and f will be less than 1 and the DQE less than ε. Figure 3.21 shows the DQE for a particular mammographic screen–film combination, taken from the work of Bunch (1999). The highest DQE value is about 0.4, somewhat smaller than the energy absorption efficiency shown in Figure 3.15. This illustrates the loss of performance due to the noise introduced as the energy absorbed from the incident x-rays is converted to film blackening. For the same reason, the DQE is quite low above about 10 lp/mm. The relatively high MTF of the mammographic screen–film system at high frequencies will not therefore provide as good a low contrast, high-frequency detection performance as might otherwise have been expected. The fall in DQE at high and low values of the photon fluence demonstrates again the limited dynamic range of the screen–film receptor.

Figure 3.21. Detective quantum efficiency for a Kodak MinR 2000 mammographic screen-film systems. The curves show the performance at 2, 4, 6, 10, 15 and 20 lp/mm. Data taken from Bunch (1999) and reproduced by kind permission of SPIE

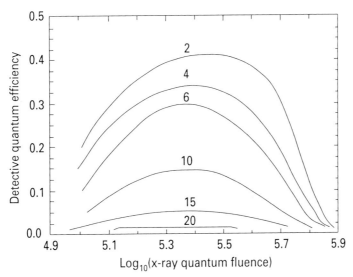

It is important to consider the actual value of SNR_{out} as well as the transfer of SNR. The noise equivalent quanta (*NEQ*) is used for this purpose. It is defined by:

$$NEQ = SNR_{out}^2 \qquad (3.5)$$

For mammographic screen–film systems the highest value of the NEQ occurs at optical densities of greater than 1.0. The desirability of using an optical density greater than 1.0 also follows from consideration of the film gamma (Figure 3.19B) and is consistent with the work of Young *et al.* (1994). These authors found that the small cancer detection rate in the UK Breast Screening Programme was dependent on the optical density on the film, with the higher detection rates being found at higher optical densities. These results are summarised in Table 3.4.

Table 3.4. Small cancer detection rates in the UK breast screening programme; data from Young *et al.* (1994)

Optical density range	Small cancer detection rate
0.80–0.99	0.13±0.02
1.00–1.19	0.11±0.01
1.20–1.39	0.16±0.01
1.40–1.59	0.17±0.01
1.60–1.79	0.18±0.02
1.80–1.99	0.20±0.02

CHAPTER 4

Digital imaging in mammography

4.1 Introduction

Various digital imaging technologies are in use for mammographic imaging. The first technology used for routine digital mammography was computed radiography (CR) (Jarlman et al., 1991; Rowlands, 2002). Small-field digital mammography devices based on image phosphors coupled to charged coupled devices (CCD) are in widespread use, primarily for stereotactic localisations (Evans DS et al., 2002; NHSBSP, 2001; Thunberg et al., 1999). Detectors for full-field digital mammography include CR which uses photostimulable phosphors. Full field direct digital radiography (DDR) systems use a number of technologies, including an image phosphor coupled to either a CCD (Smilowitz et al., 1998) or an amorphous silicon (a-Si)/thin film transistor (TFT) readout array. Amorphous selenium/thin film transistor based full-field digital mammography systems are also available.

This chapter provides an introduction to the principles and operation of these technologies. Yaffe and Rowlands (1997) have published a thorough review of digital imaging and the physics that underlies the operation of digital detectors.

The digital image is made up of an array of picture elements or pixels. The size of the pixel array and the dimensions of the individual pixels are determined by the readout device for CR systems and by the type and design of the detector for DDR systems. The highest spatial frequency that can be reproduced by a digital imaging system depends on the pixel pitch d_d (the distance between identical points on two adjacent pixels) and can be described by the Nyquist Frequency (N_v).

$$N_v = \frac{1}{2.d_d} \tag{4.1}$$

Frequencies above the Nyquist value are aliased, i.e. mirrored into the lower frequency range, thus increasing the apparent low frequency content. Aliasing may be particularly apparent when viewing images of resolution line pair test objects.

With many designs of DDR detector, the full area of each pixel may not be available to sample the signal. Figure 4.1 shows a detector element of an amorphous silicon readout device in which part of the area of the pixel is taken up by the electronic readout circuitry (thin film transistor), which is insensitive to the incoming signal. The fill factor is the ratio of the sensitive area of the pixel to the total area of the pixel.

4.2 Photostimulable phosphors

The CR image receptor comprises a layer of photostimulable phosphor deposited on a plastic substrate and is contained within a cassette similar to that for screen–film

Figure 4.1. A detector element of an amorphous silicon readout device

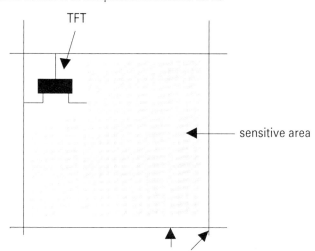

radiography. The phosphor is usually of the barium fluorohalide family (commonly BaFBr:Eu^{2+}). X-ray interactions within the phosphor result in electron excitation. Many of the electrons promptly de-excite with the subsequent release of light photons (as in a screen–film system). However, the Europium dopant produces metastable energy traps (F or colour centres) that capture the remaining electrons and a latent image of the trapped electrons is thus acquired.

A laser beam is used to stimulate the electrons from the energy traps to the conduction band and their subsequent de-excitation can result in light photons being emitted. Takahashi et al. (1984) give a thorough description of the mechanism of photostimulated luminescence.

The CR cassette is loaded into the cassette tunnel on the mammography x-ray unit and the latent image acquired on exposure. The conventional AEC system can be used to control the duration of the exposure. The cassette is then manually transferred to the CR reader in which the image plate is removed for scanning by the readout laser. The emitted light is collected by a light-guide (see Figure 4.2), and directed to a photomultiplier tube that is designed to be sensitive to the wavelength of the emitted light but relatively insensitive to scattered laser light. A filter is additionally used to remove the scattered laser light. The electronic signal from the photomultiplier tube is digitised and processed to form the image. The processing algorithm includes both field recognition and image enhancement. Field recognition is used to detect the region of the image plate that has been exposed. Further processing is applied to the data to produce an image data set that is optimised for the display mechanism being used.

Following readout, a white light source is used to clear the energy traps of residual electrons and the image plate is reloaded into the cassette. The image should be read soon after exposure to avoid a reduction in DQE due to latent image decay, which results from a reduction in the number of electrons within the metastable energy traps.

The diameter of the scanning laser beam and the spread of the emitted light limit the spatial resolution of CR imaging. A physical evaluation of a photostimulable phosphor

Figure 4.2. Reading systems for computed radiography (CR)

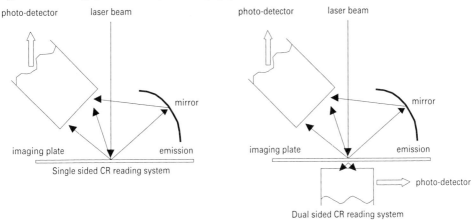

system for mammography with a pixel pitch of 100 μm has been made by Workman *et al.* (1994). A clinical comparison of this CR system with screen–film imaging has also been made by Brettle *et al.* (1994), who suggested that the image quality of the two technologies were comparable, despite the limited spatial resolution of the CR system.

A recent development in photostimulable phosphor imaging technology includes a dual read-out system. The phosphor is carried on a transparent plastic substrate allowing both sides of the image plate to be read out independently. The system has two resolution modes with a pixel pitch of either 100 μm or 50 μm. Spatial frequency dependent weighting is applied to the front and back images, which are then recombined to give the clinical image. This method should improve both spatial resolution and DQE.

4.3 Phosphor based direct digital systems

The majority of devices currently used for direct digital mammography are based on conventional image phosphors that provide the initial conversion of x-ray energy to light, which is subsequently sampled using a photosensitive detector. Phosphor materials used for digital mammography are commonly CsI or rare earth materials such as Gd_2O_2S. CsI has a columnar structure that effectively acts as a light guide to reduce the lateral spread of light and maintain sharpness.

Direct bonding of the phosphor to the readout device can be used if the dimensions of the two are the same. Alternatively, some form of demagnification device can be used to couple the phosphor to a readout device of smaller dimensions, typically a lens or lens/mirror system or a fibreoptic taper.

In Chapter 3 the importance of the position of the phosphor with respect to the image receptor is discussed. Direct digital detectors generally use a front phosphor configuration as the readout device and coupling optics could attenuate a significant proportion of the x-ray photons. It is possible to use a back phosphor configuration if a radiolucent mirror is used to couple the phosphor to a readout device positioned outside the x-ray beam.

4.3.1 Charge coupled devices

A charge coupled device (CCD) is a two dimensional electronic array that converts light into charge. It consists of a series of electrodes deposited on a semi-conductor substrate, which form an array of storage wells or pixels. Light photons from the phosphor produce charge in the semi-conductor that resides in the storage wells. Voltage switching is then used to transfer the charge from pixel to pixel during readout (Yaffe and Rowlands, 1997).

The pixel size of detectors using CCD readout can be small compared with that of detectors using other readout methods. This occurs as the fill factor for a CCD is 100%.

Figure 4.3 shows that the MTF of a phosphor system coupled to a CCD can approach that of a screen–film system.

Figure 4.3. Modulation transfer function of four digital mammography systems plotted up to the Nyquist frequencies. Curves a to c for small-field devices using a phosphor coupled to a CCD and curve d is for a full-field device using a phosphor coupled to amorphous silicon thin film transistor array. The screen-field data points are from Bunch, 1997

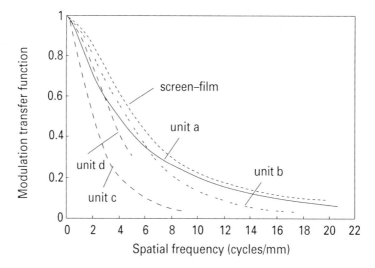

A typical CCD is approximately 2.5×2.5 cm in area. Thus a number of commercial small field digital imaging systems use demagnification optics, such as a lens or lens and mirror or a fibreoptic taper to couple the CCD to the image phosphor (typically 5×5 cm to 6×6 cm in area). As demagnification increases, the light loss increases and a spatial frequency dependent reduction in DQE may be observed (Maidment and Yaffe, 1994). This is more pronounced for systems that use lens coupling compared with those that use fibreoptics (Yaffe and Rowlands, 1997) as shown in Figure 4.4. Larger area CCDs, up to approximately 5×8 cm in area, directly coupled to the image phosphor are also used in small-field imaging systems. Larger CCDs are possible but are currently impractical due to high cost.

Figure 4.4. Effect of demagnification on collection efficiency of lens (dashed) and fibreoptic (solid) coupled imaging systems (adapted from Yaffe and Rowlands, 1997)

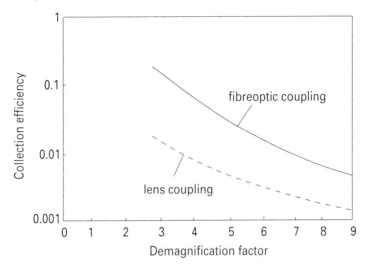

The detector housing for a unit that employs demagnification is necessarily larger than for a directly coupled device and, in use, the complete Bucky assembly is removed and replaced by the digital detector assembly. Those systems which use direct coupling are more compact and can be incorporated into a digital cassette which locates into the cassette tunnel on the existing Bucky assembly. Figure 4.5 shows the detector assemblies on two units, one that uses demagnification and the other direct coupling.

Figure 4.5. (A) Direct coupling small-field digital mammography detector assembly. (B) Fibreoptic coupling small-field digital mammography detector assembly

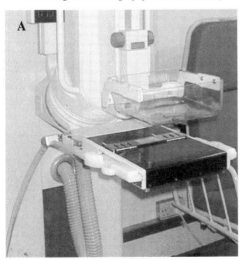 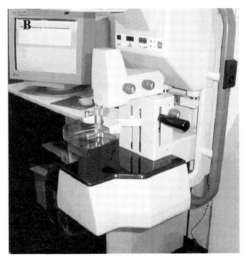

One approach for using CCDs for full-field digital mammography is to use demagnification such that a single CCD collects the light from the whole image phosphor. However, the level of demagnification can lead to a significant loss of light and a significant reduction in DQE. Another design has used an array of 12 CCDs in a 3×4 format, coupled to the image phosphor using fibreoptic tapers with a demagnification factor of approximately two. This system has a pixel pitch of 40 μm and a theoretical limiting spatial frequency (Nyquist frequency) of 12.5 lp/mm (DH, 2001b).

Imaging systems using slot beam scanning techniques are also available (Tesic et al., 1999). The detector comprises a series of long, narrow CCDs butted together and coupled to the image phosphor using fibreoptic bundles. The x-ray beam and detector assembly scan across the breast in unison. This method has the advantage of improved scatter rejection but the scanning time may be several seconds with a consequent high x-ray tube loading.

4.3.2 Amorphous silicon/thin film transistor arrays

An amorphous silicon/thin film transistor active matrix array can be used to read out the light emitted from an image phosphor (Vedantham et al., 2000). The pixels in the array each contain an amorphous silicon photodiode and thin film transistor (Figure 4.6) and are separated by readout lines. The light from the phosphor promotes charge in the photodiode, which is stored in the pixel capacitance. Each photodiode is connected to the corresponding thin film transistor and the charge read out by sequential addressing of the transistors.

Figure 4.6. Effect of light spread in phosphor based systems compared to an amorphous selenium detector

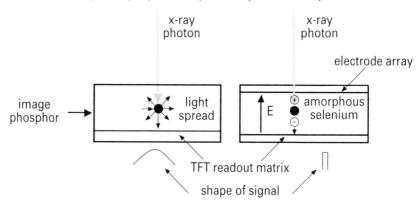

The first commercially available unit to use this technology for mammography has a pixel pitch of 100 μm. The width of the sensitive area of each pixel is approximately 87 μm and hence the fill factor is approximately 75%.

Although the limiting spatial frequency is poor in comparison to the CCD based devices and screen–film imaging, the MTF is high at the Nyquist frequency of 5 cycles/mm.

4.4 Photoconductor based direct digital systems

Another digital imaging technology in use for full-field mammography is based on a photoconductor/thin film transistor active matrix array, using amorphous selenium as the photoconductor (Fahrig *et al.*, 1995; Sexton *et al.*, 2000).

The pixel array is produced by depositing a matrix of electrodes on both sides of a layer of amorphous selenium that is coupled to a thin film transistor array. X-rays interact with the amorphous selenium, resulting in the production of electron–hole pairs. The charge is transferred to the pixel capacitance by an applied potential across the selenium layer and the thin film transistor array is sequentially addressed to read out the signal.

As the charge liberated in the photoconductor is directed by the electric field, there is almost no lateral spread of the signal. Figure 4.6 demonstrates how this aids in maintaining sharpness. Thus, although the limiting spatial frequency will be similar to phosphor based devices, the MTF is maintained at a higher level across the operating range of spatial frequencies. Also, this technology allows for the use of a relatively thick detection layer and hence has potential for higher DQE.

4.5 Comparison of digital and screen–film receptors

Conventionally, screen–film systems provide a standard image size of 18×24 cm and a large image size of 24×30 cm. CR systems for mammography provide the same nominal image sizes. Small-field digital mammography detectors are limited in field size by the dimensions of a single CCD and range from 5×5 cm to 5×8 cm. The field size for DDR systems for full-field digital mammography depends on the design and may be nominally 18×24 cm or larger. Systems providing the larger field size may allow imaging over a limited area of the field at a higher resolution.

Phosphors are used as the initial x-ray photon detector in many digital systems. The effect on spatial resolution of the lateral spread of light photons within the phosphor is discussed in Chapter 3. This, together with the finite size of the focal spot contributes to unsharpness in the image. Figure 4.3 shows the modulation transfer function for four digital mammography systems up to the Nyquist frequencies. The modulation transfer function for digital image receptors is generally inferior to that of screen–film systems.

DQE was introduced in Chapter 3 and is described as the product of the quantum detection efficiency (ε) and a factor (f) that relates to various loss and noise processes in the imaging system. Noise sources, which contribute to a reduction in the value of f, differ for each technology but include secondary quantum noise for phosphor based systems, secondary charge carrier noise, thermal noise and losses due to optical coupling. DQE is considered a more complete descriptor of imaging performance than MTF (Cowen *et al.*, 1997).

Screen–film systems can have a peak DQE of up to 40% at low frequencies but maintain a higher DQE at high spatial frequencies (Bunch, 1997). Due to the shape of the characteristic curve and limited latitude of the screen–film systems, the DQE is significantly reduced at higher and lower exposure levels. This is due to the effect of the exposure dependent contrast transfer function on the relative magnitude of the quantum noise. Additionally, a reduction in DQE can be observed due to screen structure noise and film granularity.

Digital detectors have a linear response allowing them to maintain their peak DQE across a wide exposure range. The DQE for a digital detector, shown in Figure 4.7, can be compared with that for a screen–film system shown in Chapter 3. Flat field calibration is used with digital image receptors to remove structural noise from the image. Dark noise subtraction is used to reduce the contribution of thermal noise.

Figure 4.7. Detective quantum efficiency (DQE) of a full-field digital mammography system that uses a phosphor with an amorphous silicon thin film transistor readout device

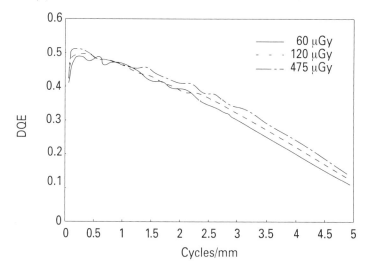

Insensitive areas in digital detectors must be kept to a minimum to enable as many x-ray photons as possible to contribute to image formation. Such areas result, for example, at the joints where more than one detector or readout device is used to cover the image area. Thin film transistors are insensitive to x-rays and detectors that use such devices have a reduced geometric efficiency as described by the fill factor (Figure 4.1). Reducing the size of the pixel to improve spatial resolution results in a reduction of the fill factor (with a consequent reduction in detection sensitivity) as the area of the thin film transistor cannot be reduced. This problem does not occur with screen–film, photostimulable phosphors and detectors using phosphors coupled to a CCD as these systems have a fill factor of 100%.

4.6 Image display and digital processing

Digital mammograms can be viewed either on a workstation monitor (softcopy) or by printing on to film (hardcopy).

For full field digital imaging the acquisition workstation allows input and review of patient and examination information and provides a limited range of image processing facilities. However, the monitor will generally not be of a high enough specification to be suitable for reporting. A reporting workstation is typically provided with two high specification, high resolution monitors mounted adjacent to each other in portrait orientation. The workstation carries a comprehensive range of image processing tools to allow image manipulation, plus image handling, storage and communication facilities. Small field digital systems are generally provided with a single workstation designed for acquisition and image review.

High specification workstation monitors are available with a resolution that can adequately display most digital mammograms. However, depending on the pixel size of the image, zoom or magnification may need to be used to display the image at full resolution. Additionally, the dynamic range of the monitor may be insufficient to adequately display the full contrast range in the image. Thus there is some concern that digital mammography may lead to an increase in the reporting time required to allow for optimal adjustment of the image. Ideally, the workstation should be provided with a range of processing algorithms designed to produce optimal softcopy images for the intensity range imaged. Further time may be required if options such as computer aided detection (CAD) are used.

Figure 4.8. Reporting workstation for digital mammography with two high quality monitors for viewing pairs of mammograms (courtesy of GE Medical Systems)

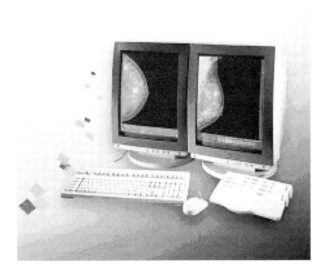

Laser printers for hardcopy images are available with adequate resolution (down to a pixel size of approximately 40 µm) and adequate dynamic range (optical density range of 0.5–3.5) that are sufficient to display the contrast range within the mammogram. The printer must be set up to give a true representation of softcopy image.

4.7 Artefacts

(i) Dead pixels

For systems other than CR, in the normal image presentation format with the greyscale inverted, a single dead pixel will appear as a white dot. A cluster of neighbouring dead pixels within a region can appear similar to a micro-calcification. Some systems are able monitor the pixel matrix for dead pixels and interpolate a value from the neighbouring pixels (Thunberg et al., 1999). Dead pixels can also result from scratches or blemishes on a photostimulable phosphor plate.

(ii) Line drop-out

A damaged readout line for digital systems other than CR will appear as a dead line in the image. The same effect can result from incorrect function of the scanning laser for CR systems.

(iii) Dust artefacts

Dust on the image plate of a CR system appears as white dots and could be mistaken for micro-calcifications. The image plate can be cleaned using appropriate cleaning materials. Dust particles on the collection optics of a CR system can appear as a dead line in the image.

Dust on the lens or mirror of a phosphor system optically coupled to a CCD readout device appears as white dots and, again, could be mistaken for micro-calcifications.

(iv) Stitching lines

Stitching lines result where more than one detector or readout device is used. This is due to the insensitive region where the detectors or readout devices meet. Although stitching lines are small (typically 80–100 µm in width), manufacturers may use image processing algorithms to minimise their appearance.

(v) Checker board effect

A checker board appearance can result when using more than one detector or readout device due to variations in thermal response. This should not arise if the detector assembly is maintained at or close to the temperature at which it was calibrated.

(vi) Ghosting

A ghost of the previous image may appear with a CR image plate if the residual image that remains after readout is not fully removed during the erasure cycle. Afterglow on other digital systems can also result in ghosting. Therefore, when testing digital imaging systems, it is advisable to protect the detector from repeated high exposures with a suitable attenuator to prevent the possibility of burning an image onto the receptor.

CHAPTER 5

The mammographic x-ray unit

5.1 Introduction

Modern mammography units are designed to allow easy examination of the breast. They have a dedicated high voltage generator to provide the required range of tube voltage. In view of the requirements for high output and short exposure times and near constant tube voltage, a medium/high frequency inverter generator is normally provided. The x-ray tube/collimator assembly and breast support platform are mounted on a U-arm that can be moved vertically on a support column. The U-arm can also be rotated to provide various projections.

Until recently, the preferred target/filter combination for screen–film imaging was molybdenum/molybdenum, as this combination gives an acceptable balance between dose and image quality for an average breast. Most current mammography x-ray units provide the option of alternative filter materials that include rhodium and aluminium in addition to molybdenum. X-ray tubes with dual target tracks (molybdenum and rhodium or molybdenum and tungsten) are also in use. By selecting an appropriate combination of target and filter materials and tube voltage, it should be possible to optimise the spectrum for different types of breast.

The nominal focal spot size of a modern mammography x-ray unit is typically 0.3 mm. If a magnification facility is provided, a dual focus x-ray tube is required with an additional fine focus, typically of 0.1 mm. The window of the x-ray tube is generally beryllium; the inherent filtration of the x-ray tube and shield should not exceed 1 mm beryllium (DH, 2001a).

The breast support platform incorporates a cassette tunnel, a moving anti-scatter radiation grid and an automatic exposure control (AEC) detector. An alternative cassette holder for direct exposure without the grid and a magnification platform may also be provided. Modern mammography cassettes are constructed from hard plastic or carbon fibre materials (both of low attenuation) and have a single, back intensifying screen for use with single emulsion film.

One of the most important factors when reviewing mammograms is for the images to be correctly exposed and uniformly presented. To achieve this an optical density in the range 1.5–1.9 has been recommended (NHSBSP, 2004). In order to maintain the optimum density for the range of breast sizes and tissue densities encountered, a reliable and consistent AEC system is essential. The AEC detector should be movable, to allow for imaging different sizes of breast and areas of the breast. The detector is located after the image receptor and monitors the quantity of x-rays transmitted through the breast and cassette. Exposure is terminated when the integrated signal reaches a predetermined value that will provide the required optical density on the processed film.

It is important that the response of AEC systems is closely matched to the speed and energy characteristics of the screen–film combination in use. It should also compensate for changes in exposure rate due to differences in breast thickness or composition, changes in tube voltage and target/filter combination and reciprocity failure. The detector may be an ionisation chamber, photo-detector or solid-state device. Operation should be possible in magnification mode and with a stereotactic device, if provided. A post-exposure readout of delivered mAs must be provided to allow calculation of breast dose.

Beam optimisation software can operate in a number of ways. On some units the exposure parameters can be determined via the AEC system. A brief initial exposure is made which assesses the attenuation of the breast. This, in combination with the compressed breast thickness, is used to determine the optimum target material, filter material and tube voltage. The initial exposure is closely followed by the actual exposure. In some other systems the exposure parameters are determined solely from the compressed breast thickness. Alternatively, the tube voltage and, in some cases, the filter may be rapidly changed during the first part of the exposure depending on the exposure rate measured at the AEC detector.

Breast compression is essential in mammography. The compression drive must be motorised and the device should be designed with smooth surfaces and without sharp corners or edges. It is important to provide the option of automatic release of compression after exposure and an emergency release in case of system failure. To prevent the possibility of excess compression, it must be possible to set a maximum level that cannot be exceeded.

Many of the measurements described in Section 5.6 need to be carried out under standard exposure conditions. Unless otherwise specified, it is suggested that this should be 28 kV, broad focus using a molybdenum/molybdenum target/filter combination. It is important that instrumentation used for these measurements is calibrated. Recommended frequencies of tests are given in Appendix I.

On completion of any test procedures, care must be taken to ensure that all test equipment is removed from the x-ray unit and that all controls are returned to the original settings. The compression plate and face guard should be re-fitted and the AEC detector returned to the original position.

Although the tests required for small and full-field digital mammography systems are generally similar to those for screen–film imaging, the methodology may differ in some cases (NHSBSP, 2001). This is discussed in this chapter where appropriate.

The Department of Health publishes a series of NHSBSP Guidance Notes intended to assist prospective purchasers in drawing up their equipment purchasing specification. The current edition of the Equipment Guidance Notes (DH, 2001a) provides more information relating to the design and performance of mammography x-ray units.

5.2 Electrical safety

The major requirements for electrical safety are contained in IEC 601-1, 1988 (BS 5724: Pt 1, 1989) 'Medical Electrical Equipment, General Requirements for Safety' and in Department of Health document TRS89, 'Technical Requirements for the Supply and Installation of Apparatus for Diagnostic Imaging and Radiotherapy' (DH, 1989c).

The responsibility for initial testing lies with the supplier. However, a representative of the purchaser should normally make further checks at acceptance. This may be an x-ray engineer employed by the hospital, a consulting engineer or a medical physicist. It is important that these checks are made and that the line of responsibility is clearly laid down (NHSBSP, 2003b). The medical physicist may not wish to perform these tests and may need to formally disclaim responsibility if insufficient training and experience in this field has been provided. Further details covering the scope and methods of electrical safety testing may be found in DHSS (1985), DH (1989b) and NHSBSP (2003b).

5.3 Mechanical safety and function

5.3.1 Introduction

The mechanical safety aspects of dedicated mammography equipment differ in some aspects from those of other x-ray units. A compression device is fitted and on most modern units is power driven although a manual override may be provided. There may also be power-driven height adjustment and rotation of the of the U-arm assembly. A radiation protection screen often forms part of the equipment. Many of the following checks may need to be repeated regularly to ensure continuing safe operation; the frequency will vary depending on the function tested (DH, 1989c, 2001a; IEC, 1988/BS, 1989; IPEM, 2002; NHSBSP, 2003b).

5.3.2 Mechanical safety checks

The following features of the equipment should be checked:

(i) Manual and powered movement of the U-arm should be prevented when compression is applied.

(ii) The automatic release of the compression plate after an exposure should function correctly (if fitted). The override of automatic release should also function correctly.

(iii) Provision for emergency release of compression, in the event of system failure, must be provided and must operate effectively.

(iv) For power driven compression, the compression device must not be able to apply a force exceeding 200 N.

(v) There should be no sharp edges or surfaces on the compression plates, breast support platform, etc. which may injure the patient.

(vi) The edges of the radiation protection screen should be clearly defined so that the operator is aware of the outline.

(vii) The restraining devices provided on mobile units (trailers and caravans) to prevent damage to x-ray unit, radiation protection screen, etc. during transit should be effective in use.

Additional checks can include a visual inspection of the unit to ensure that no parts are damaged. Compression plates should be examined for possible cracks (these may only be apparent under compression) and sharp edges.

5.3.3 Marking and labelling

The following should be clearly marked or indicated:

(i) The focal spot size and position.

(ii) The amount of inherent, added and total filtration (usually in mm of aluminium) including that of alterable or removable filters.

(iii) The position of AEC detectors.

(iv) The magnification factor(s) of any magnification device. Note that the specified magnification factor for different equipment manufacturers may refer to different reference planes (e.g. on the magnification platform or 4 cm above the magnification platform).

(v) The function of all controls.

5.3.4 Mechanical function checks

The following features of the equipment should be checked:

(i) All manually controlled movements should operate smoothly and be free running, and the force required to be exerted by the operator to cause any movement should be less than 30 N.

(ii) All mechanical/electromechanical brakes should function correctly and without backlash.

(iii) All scales/indications on linear/rotational movements and the indication of focus film distance (FFD) if adjustable, should be clearly marked.

(iv) All beam limiting diaphragms must be marked with their field sizes at the relevant FFD.

(v) Power driven vertical movement should be possible with the patient leaning against the breast support platform (without compression applied).

(vi) Vertical and rotational movement, whether manual or power driven, must be prevented when compression is applied.

(vii) All foot switches should operate correctly.

(viii) All attachments should locate correctly and their locks should function effectively.

(ix) The AEC detector should locate effectively into the pre-set positions.

(x) The Bucky assembly should provide firm retention of the cassette (with the U-arm both vertical and horizontal) but allow easy insertion and removal.

(xi) The interlock to prevent exposure when the cassette is not correctly positioned should operate correctly.

(xii) The light intensity from the x-ray field light should be adequate (DH, 2001a).

(xiii) The movement of the compression carriage should be smooth.

(xiv) The indication of breast thickness must be accurate.

5.4 Radiation safety

5.4.1 Introduction

Radiation safety requirements are defined by the Ionising Radiations Regulations 1999 (HMSO, 1999), the Approved Code of Practice and Guidance (HSE, 2000), the Medical and Dental Guidance Notes (IPEM, 2002) and the Ionising Radiation (Medical Exposure) Regulations 2000 (HMSO, 2000). The TRS89 (DH, 1989c) also gives some guidance.

The differences between mammography units and general x-ray units, as far as radiation safety is concerned, arise from the use of lower x-ray energies and the specialised geometry. The x-ray field is permanently aligned with the breast support platform and should be only slightly larger than the area of the image receptor. The breast support platform should also act as a primary beam absorber.

X-ray beam alignment is particularly important. Ideally, the beam should just cover the film. In particular, the x-ray field should extend to the film edge closest to the chest wall and may extend slightly beyond it in order to ensure that this region of the breast is fully imaged. Some mammography x-ray units provide automatic or motorised collimation; others have a series of interchangeable beam limiting diaphragms for different applications.

5.4.2 Inspection

Regulation 32 of the Ionising Radiations Regulations 1999 (HMSO, 1999) requires that x-ray equipment is designed or constructed and installed or maintained such that the amount of radiation received by patients is the minimum consistent with the clinical requirements. Additionally, Regulation 32 states that x-ray equipment should, where practicable, be provided with suitable means of indicating the amount of radiation delivered during a procedure. On most mammography systems this is satisfied by an indication of the delivered mAs, although certain systems can provide a direct indication of the breast dose.

The following features should be checked and reference should be made to IPEM (2002), DH (1989c) and DH (2001a):

(i) A mains isolator, accessible from the normal operating position should be provided.

(ii) A visible indication must be provided on the control panel to show that the mains is switched on.

(iii) The visible exposure warning indication must function correctly and remain 'on' long enough to be seen even at the shortest exposure times. An audible warning may also be provided but is not a substitute for a visible warning.

(iv) The total filtration must be equivalent to at least 0.5 mm Al or 0.03 mm Mo.

(v) If the added filtration is removable or interchangeable, an interlock must be provided to prevent exposure if the filter is removed or incorrectly inserted.

(vi) If the field-limiting diaphragm can be removed, an interlock should be provided to prevent exposure unless the diaphragm is properly aligned.

(vii) The exposure must terminate if the exposure control is released prematurely.

(viii) The location of the exposure control should confine the operator to the protected area during exposure.

(ix) The exposure control should be designed to prevent inadvertent production of x-rays.

(x) The design of the exposure control should prevent further exposure unless pressure on the control is first released.

5.4.3 Integral radiation protection screen

A radiation protection screen should be provided to afford protection equivalent to at least 0.1 mm of lead at 50 kV and should allow good visibility of the patient by the operator and vice versa (DH, 2001a).

The lead equivalence of the radiation protection screen should be marked (on both the glass and the panel where appropriate) at a specified voltage (DH, 1989c; IPEM, 2002). If the lead equivalence is not marked and is not shown in the accompanying documentation, it will need to be measured. Measurement of lead equivalence is not straightforward but it may be deduced from the measurement of the transmission of primary or scattered x-rays through the radiation protection screen. Alternatively, the radiation from a suitable low energy radionuclide source may be used for this measurement.

5.4.4   X-ray room

A room warning notice and a visible x-ray warning signal should be provided at all entrances to the x-ray room. These should indicate when x-rays are being or are about to be generated (IPEM, 2002).

A check on the room shielding, either visually, against the Radiation Protection Adviser's requirements at the planning stage, or by transmission measurements, should be undertaken at or prior to installation. Walker and Hounsell (1989) and Sutton and Williams (2000) give data on appropriate shielding for walls and ceilings of mammographic x-ray rooms. To check environmental radiation levels under realistic conditions, film badges, thermoluminescent dosimeters or other suitable dose monitors may be positioned in the x-ray room.

5.5   Additional checks for mobile equipment

Mobile mammography units (trailers or caravans) are subject to considerable vibrations during transit. Additionally, the mechanical rigidity of the unit may be such that vibration may be experienced due to windy conditions or to persons moving within the vehicle during exposures. Changes in line voltage or supply impedance between sites (if local supplies are used) may affect the operation and calibration of the x-ray generator.

For these reasons the following checks should be considered when a vehicle is moved between sites.

5.5.1   Before moving

It is normally left to radiographers or other staff operating the mobile unit to ensure that all freely moveable objects/equipment are firmly locked or strapped so that damage during transit is avoided. This will apply to items such as film loaders, chairs or mobile radiation protection screens. The x-ray unit may be provided with restraining devices, which should be used to prevent undue strain on the vertical and rotational movements of the U-arm or damage to x-ray tube or radiation protection screen during transit.

5.5.2   After moving

Ensure that all restraining devices are removed and perform the following checks:

(i)   The breast support platform and associated equipment should be visually checked for possible damage.

(ii)   The mechanical function and safety aspects of the compression device should be checked.

(iii)   The alignment of the x-ray beam to the image receptor should be checked.

(iv)   Repeatability of AEC system and constancy with change in phantom thickness should be tested.

(v)   Image quality with a standard phantom should be tested.

The radiographer would normally make these checks on-site and would contact the medical physicist if the results were outside reference values stated in the quality assurance manual. Depending on the results, it may be necessary for the medical physicist to check accuracy of tube voltage, radiation output and/or focal spot size.

5.6   Tests on the mammographic x-ray unit

5.6.1   Introduction

The tests listed in this section are specific to the mammographic x-ray unit. Although they are described in full, some tests are only performed on commissioning or when testing a new x-ray tube. A subset is performed on a routine basis. The range of measurements made routinely may be reduced to test those parameters used most frequently in the clinical situation. A list of suggested test frequencies is given in Appendix I.

5.6.2   X-ray tube rating

Before commencing any measurements, it is important to refer to the x-ray tube data to find a suitable exposure repetition rate that is within the rating of the x-ray tube. Most modern x-ray tubes will allow exposures in excess of 100 mAs per minute, but 50 mAs per minute may be a more realistic rate to avoid tube damage. Some mammography units provide a delay period after each exposure during which further exposure is prevented. This may be a set time interval (typically 30 seconds) or an interval derived from the parameters and frequency of the previous exposures.

5.6.3   Alignment

With care, it is possible to minimise the number of films needed to check the alignment. Although for some of these tests a fluorescent screen may be used in place of film, a film method has the advantage of providing a permanent record. If, when checking the alignment of the light field to the radiation field, metal markers are used to outline the light field, the alignment of the radiation field to the film in the cassette and the separation between the film edge and the edge of the breast support platform can be checked at the same time.

If a 24×30 cm Bucky assembly is provided and is used clinically, the tests described in paragraphs (i), (ii), and (iii) should be repeated. If a magnification device is provided and is used clinically, these tests should be repeated for the magnification mode using fine focus.

(i) Alignment of light field to x-ray field

At commissioning, the alignment of the light field to the radiation field should be checked for all targets with all the combinations of focus to film distance (FFD) and field size available for contact and magnification films. For routine testing, if fewer FFD and field sizes are used clinically, the number of tests can be reduced. Position a loaded screen–film cassette on top of the breast support platform, place markers on the edges of the light field and additionally in one corner of the field to indicate the orientation. Make an exposure and measure the displacement between the light field and x-ray field along the four field edges on the processed film.

Equipment: screen–film cassette, markers (e.g. stiff wire, coins, paper clips) and steel rule

Remedial: misalignment of light field to x-ray field >5 mm along any edge

(ii) Alignment of x-ray field to film

The alignment of the x-ray field to the film in the cassette should be such that the whole of the film is exposed but with minimal overlap. The tolerance along the chest wall edge is particularly critical. Ideally the edge of the x-ray beam should lie between the edge of the film and the chest wall edge of the breast support platform (assuming that the combined thickness of the edge of the cassette and front of platform will be of the order of 5 mm). This is to ensure adequate coverage of tissue close to the chest wall. The alignment of the x-ray beam to the rear (nipple) edge may be thought to be less critical. However, an unexposed area of film cut-off here (either by collimator, cassette holder or the nipple edge of the compression device) can result in glare when viewing the processed films or may increase the number of women who require additional films to image the whole breast. The design of the x-ray set may result in cut-off that cannot be corrected. This is usually at the nipple edge. Cut-off on any of the other edges can usually be corrected.

The accuracy of alignment should be checked for all the combinations of FFD and field size available and for both contact and magnification films (with cassette holder if applicable) for all targets and foci. There are several methods of accurately checking alignment (Brennan and Johnson, 1993; Kotre *et al.*, 1993) using specially designed test tools. Alternatively, a loaded mammography cassette may be placed in the normal position in the cassette tunnel and a larger 24×30 cm cassette (or two 18×24 cm cassettes at 90° to their usual orientation) placed on the breast support platform so that it/they overlap the edge of the platform.

The platform edge should be marked with a piece of stiff wire (e.g. welding rod), and other markers (e.g. coins) are placed in other positions so that the orientation of the films can be determined. The identity and orientation of films should be labelled. After making an exposure, the displacement between the x-ray field and the film edges is estimated from the two processed films by aligning them using the positions of the markers (allowing for magnification if necessary). Care should be taken to ensure that the optical density of the films allows the penumbra to be visualised. When applying the limits below, it may be necessary to take into consideration movement of the film within the cassette, as well as movement of the cassette inside the Bucky.

Figure 5.1. Plan view of equipment layout for alignment tests

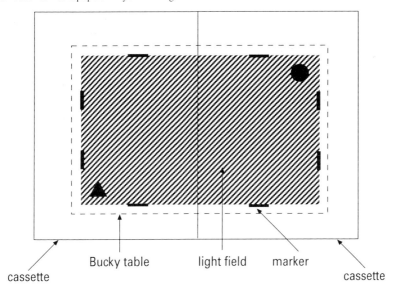

Equipment: screen–film cassette, markers (e.g. stiff wire, coins, paper clips), and steel rule

Remedial: >5 mm or <0 mm overlap of film by x-ray field along all sides

Suspension: >10 mm overlap or >2 mm unexposed border along chest wall edge with respect to the film

>10 mm overlap along left or right edge with respect to the film

(iii)   Separation between film edge and edge of the breast support platform

The front edge of the film in the cassette should be as close as possible to the front edge of the breast support platform so that as much of the breast as possible is imaged. The distance from the edge of the film to the front edge of the platform, for the normal position of the cassette, should therefore be checked.

Equipment: screen–film cassette, markers (e.g. stiff wire, coins, paper clips), and steel rule

Remedial: >5 mm between edge of film and the front edge of the breast support platform

(iv)   Alignment checks for digital systems

For digital systems, the alignment of the x-ray field to the detector and to the light field can be checked using a similar method as for conventional systems. However, the recorded digital image will replace the film from the cassette in the Bucky assembly. Tolerances are the same as for screen–film systems.

(v)   Size of imaged field for digital mammography

This test is performed to verify the nominal size of the image field. It requires knowledge of the alignment of the light beam to the detector (see above). Position suitable markers or a radio-opaque scale as close to the detector plane as achievable (usually on the breast support platform) such that they delineate an area approximating the nominal detector size. A geometrical correction may be required due to the markers being positioned on a plane above that of the detector. Make an exposure and check that the markers are visible on the image at the workstation monitor. Where provided, the hardcopy image should be tested independently. If using callipers to determine the image size, the accuracy of these at the same plane must be verified prior to performing this test.

Remedial: either dimension 5% less than stated nominal size

5.6.4   Leakage radiation

It is very unusual for a modern mammographic x-ray tube assembly to develop defects in the tube shielding. However, in view of the proximity of the tube housing to the woman being examined, it is important that possibility of leakage should be checked.

The Medical and Dental Guidance Notes (IPEM, 2002) suggest that the detector used for this measurement should have an effective cross-sectional area that enables the leakage to be averaged over an area not exceeding 100 cm^2. However, this refers to a distance of 1 m from the tube focus and for mammography x-ray tubes detection of any leakage at this distance is unlikely. Therefore, practically, checks are usually first made by using an ionisation chamber positioned on or very close (5 cm) to the tube housing. Exposures should be made at several positions around the tube head at the maximum tube voltage and typically 100 mAs. For a typical mammographic x-ray tube any reading greater than 10 µGy (with the chamber in this position) should be further investigated. The readings should be corrected to mGy in 1 hour at 1 m from the focus at the maximum rating of the x-ray tube. Any deficiencies or defects in the shielding of the tube housing or diaphragm assembly can best be further pinpointed using film (ideally envelope wrapped film or film in flexible cassettes). For all of these tests a beam-stopper of at least 1 mm lead should be placed over the collimator port such that no primary radiation is emitted.

Equipment: ionisation chamber system
pre-packed film/industrial flexible cassettes/ordinary cassettes fitted with intensifying screens
x-ray tube rating charts

Suspension level: leakage radiation >1 mGy in 1 hour at 1 m from the focus at the maximum rating of the tube assembly averaged over an area not exceeding 100 cm^2 (IPEM, 2002)

5.6.5 Compression and breast thickness indication

(a) Introduction

The design and operation of the compression device must ensure that the whole of the breast is firmly compressed and securely held, and should cater for all sizes and types of breast. Compression plates are available, constructed using flexible and rigid plastics; some are designed to tilt toward the chest wall or nipple edges on compression. Foot switches are normally used to control the movement of the compression plate; movement should only occur when the switch is depressed. The accuracy of the thickness indicator is of importance if the automatic exposure control system uses beam quality optimisation software based on breast thickness. The calibration of the indication must also be known in order to calculate the mean glandular dose correctly.

(b) Measurements

(i) Compression force

The compression force can be measured using a force balance, strain gauge or flat faced set of scales (such as bathroom scales). The device should have a continuous readout and should be regularly calibrated. A compressible object (e.g. hard compressible foam) should be placed on the measuring device in order to distribute the load across the compression plate (similar to the clinical situation). The maximum compression force for both power driven and manual application and the accuracy of the displayed compression force (across the range of force used clinically) should be determined. To ensure that the compression force is maintained throughout the exposure, any change in force over a 30 s period after initial application should be noted.

Equipment:	force balance, strain gauge or scales
	compressible object
Remedial:	difference between measured compression force and indicated compression force >20 N
Suspension:	maximum power driven compression force <150 N or >200 N (and cannot be brought within this range by adjustment of a user accessible control)
	maximum compression force in any mode of operation >300 N
	>20 N change in compression force over 30 s period

(ii) Indication of thickness

Slabs of Perspex 2–8 cm thick should be aligned with the chest wall edge of the breast support platform. The compression plate should be in contact with the Perspex and a compression force of between 50 and 100 N should be applied. (The applied force for this measurement should be established and the same force used for all subsequent measurements.) Ideally 18×24 cm slabs of Perspex should be used to prevent deformation of the plate and reduce measurement inaccuracies on the indication of thickness due to the tilt angle of the plate. If semi-circular Perspex slabs are used, they may be aligned perpendicular to the chest wall to minimise both of these effects. The distance from the compression plate to the breast support platform (thickness of the Perspex slabs) should be measured on the midline at the chest wall and compared with the displayed value. It may be necessary to repeat the measurements in magnification mode.

Equipment: Perspex slabs of known thickness

steel rule

Remedial: difference between measured thickness and indicated thickness >5 mm

Suspension: difference between measured thickness and indicated thickness >10 mm if AEC operation is affected by scale inaccuracies

5.6.6   Dimensions of focal spot

(a)   Introduction

IEC 336, 1993a (British Standard 6530, 1994) describes three techniques that may be used to measure the characteristics of the focal spot. Although this Standard refers in general to conventional diagnostic x-ray tubes, much of the information may be applied to mammographic units.

The Standard recommends that the focal spot dimensions are measured using a slit camera but recognises that a star resolution grid and a pinhole camera are also useful. The results given by these three techniques may differ. Kimme-Smith *et al.* (1988) have compared the slit and pinhole methods, and Doi *et al.* (1982) and Everson and Gray (1987) have compared all three. Since these publications, the multiple pinhole test tool has also been used for measuring focal spot dimensions (Jacobson, 1994).

It is suggested that on installation of a mammography system, the dimensions of the focal spot should be checked using the slit camera method (IEC, 1993a). This method requires use of a jig (e.g. Ramsdale *et al.*, 1989) for accurate alignment, as does the pinhole method. An image of a pinhole or a star resolution grid may be recorded on the same occasion, as the shape and density distribution of the image may give a valuable indication of tube life. Subsequently, as part of the routine tests, further images can demonstrate any appreciable change in the focal spot. There may be difficulties in the interpretation of star pattern images (Burgess, 1977; IEC, 1993a; Spiegler and Breckinridge, 1972).

The recommended tube voltage for focal spot measurements in IEC 336 is 75 kV, which is not suitable for mammography x-ray units. Therefore the test should be performed at a tube voltage of 28 kV. Measurements should be made for each focus and each target. Care should be taken to use the correct exposure as this can affect the apparent focal spot size. If the image is too dark, the edges of the focal spot cannot be easily discriminated.

(b)   Measurement geometry

A suitable support should be constructed to position the slit camera, star resolution grid or pinhole between the tube focus and the breast support platform, close to the collimator port. For reproducible and accurate measurement of the dimensions of the focal spot, it is important that the measuring device is aligned along a specified reference axis. This may or may not be the central beam axis and may or may not be normal to the platform

(IEC, 1993a). For routine work, the axes suggested in DH (2001a) that meet the breast support platform at 5 cm from the chest wall for broad focus and 2 cm for fine focus can be used. A second measurement on the reference axis quoted by the manufacturer may be necessary if the user wishes to show that the focal spot is not within the manufacturer's specification. In practice it may be more convenient, particularly with the star pattern, to measure the effective focal spot dimensions at some other angle (q) and then correct the measured length (l_ω) to the required reference axis using the relation:

$$l_\omega = l[\tan \phi - \tan \omega]/[\tan \phi - \tan \theta] \tag{5.1}$$

where ϕ is the target angle with respect to the normal to the film plane (i.e. will generally be the nominal target angle *plus* any tube angle) and ω is the 'reference angle' specified by the manufacturer (see Figure 5.2). This equation only strictly applies when the film is placed flat on the breast support platform (e.g. in making a measurement with the star pattern – see Section c(iv)). If the film is aligned at right angles to the reference axis then an additional error will be introduced, but this will be small for measurement angles of 10° or less. The measuring device should be located close to the tube port to achieve adequate magnification. This may require the removal of a collimator cone or face guard, if fitted. The magnification may be determined geometrically or radiographically. Data may be obtained from the x-ray tube data or from the manufacturer (BSI, 1981).

Figure 5.2. Focal spot geometry

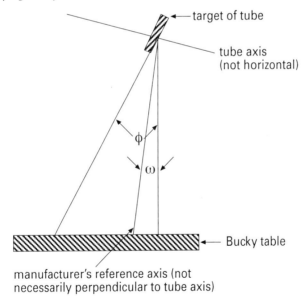

The nominal focal spot value specified by the manufacturer will generally be smaller than the measured dimensions due to the generous tolerances allowed in IEC 336 (1993a). In the following paragraphs the term 'nominal focal spot value' refers to that quoted by the manufacturer. It is important that the measuring method used by the manufacturer is known when comparing measured dimensions to quoted values.

(c)   Measuring methods

The methods for testing the focal spot size rely either on examining the shape of the focal spot (slit, pinhole and multi-pinhole) or the degradation in resolution (star pattern).

(i)   Slit camera

The focal spot dimensions should be determined using a slit camera with a 10 μm slit. Two magnified images should be produced with the slit normal to and parallel to the x-ray tube anode–cathode axis. The measuring geometry should be such that the magnification is as large as practicable, typically 2.5–3×. The film should be exposed such that the optical density of the image of the slit is between 0.8 and 1.2 above the base plus fog level. The use of a standard mammographic screen–film combination is acceptable and gives no significant difference in focal spot size measurements when compared with non-screen film such as dental film. Users should note that mechanical vibration, for example on a mobile breast screening unit, could affect these measurements, particularly with fine focus. It may be useful to make the measurement under conditions of clinical use, including vibration, if normally present. However, if a comparison with the manufacturer's specification is being made, vibration should be excluded.

The dimensions of the focal spot are derived by examining and measuring the pair of images through a magnifier (typically at least 10×, having a built-in graticule with 0.1 mm divisions) and correcting for magnification. Alternatively, a scanning densitometer or travelling microscope can be used to measure the dimensions of the images. If the magnification is not great enough, the precision of the measurement can suffer.

If the magnification factor (M) is taken as the ratio of the size of the slit camera support on the image to the actual size, the dimension of the focal spot (f) is given by:

$$f = d/(M-1) \tag{5.2}$$

where d is the size measured on the image.

(ii)   Pinhole

The pinhole method is similar to the slit camera method and produces an image of the focal spot that shows the intensity distribution across the focus. In addition to the dimensions, the image may give information relating to the condition of the focal area. A gold/platinum alloy disc with a pinhole 30 μm in diameter is supported in a similar manner to the slit camera. This size of pinhole is only suitable for measurement of broad focus as pinholes with a small enough diameter for fine focus (7–10 μm) are not readily available. The image may be recorded using non-screen film or a standard mammographic screen–film combination. The correction for magnification is as for the slit camera.

(iii)   Multi-pinhole

The multi-pinhole (a stainless steel plate with a regular array of pinholes, 50 μm in diameter) produces multiple images of the focal spot. As with the single pinhole technique the image shows the intensity distribution across the focus and may show its condition in addition to giving dimensional information. The plate is attached to the underside of the x-ray collimator and the image may be recorded using non-screen film or a standard mammographic screen–film combination. As multiple images are produced, positioning is not as critical as for the single pinhole method. The size of the pinhole image can be corrected to the required reference axis by interpolation. Due to the size of the individual pinholes (50 μm), this technique is only suitable for measurement of broad focus. As with the earlier methods, the magnification factor can be determined using the film produced and the known spacing of the pinholes.

(iv)   Star resolution grid/sector grid

The focal spot dimensions can be estimated from the 'blurring diameter' on the image of a star resolution grid (star pattern). This diameter refers to the distance between the outermost blurred regions on the image along each direction of evaluation (normal to and parallel to the x-ray tube anode–cathode axis). It should be noted that the blurring diameter measured across the tube axis gives the effective *length* of the focus whilst that measured parallel to the tube axis gives the effective *width*. It is often found that the former diameter is difficult (or sometimes impossible) to identify with small foci.

Suitable spoke angles are 1° or 0.5° for focal spot sizes >0.2 mm and 0.5° for focal spot sizes <0.2 mm. For focal spot sizes of approximately 0.1 mm, a tapered line-pair bar pattern (sector grid) up to 20 lp/mm may be more useful. Exposures are made with the pattern arranged across and along the tube axis to obtain focal length and width respectively.

The star pattern is attached to the underside of the x-ray collimator and is centred using the light beam. By marking the front edge of the platform with wire the measurement angle (q) may be calculated from the resulting film and the estimated length corrected to the value on the reference axis (1_ω) as previously described (Section 5.6.6(b)). A magnified image of the star pattern or bar pattern is produced on non-screen film. Alternatively, a standard mammographic screen–film combination can be used with approximately 2.5 mm Al in the beam (e.g. on top of the star pattern) to achieve the correct exposure.

For a star pattern with a spoke angle of $\theta°$ the effective dimension f is given by:

$$f = \pi\theta \times D/180(M-1) \tag{5.3}$$

where D is the blurring diameter on the image and M is the magnification factor. M is the ratio of the diameter of the star pattern on the image to the actual size.

(d)   Results

Results of measurements in the UK by slit and star pattern methods are given by Young *et al.* (1992). The measured values should be within the manufacturer's specification and it is important that the measuring method quoted by the manufacturer is taken into account.

It should be noted that manufacturers quoting focal spot sizes in accordance with the IEC (1993a) standards generally apply a +50% tolerance. For nominal focal spot values >0.3 mm an additional factor of 0.7 is applied to the focal length. Thus a nominal 0.3 mm focus may be up to 0.65 mm long × 0.45 mm wide on the reference axis. It is not certain whether the factor of 0.7 is intended to apply to mammographic x-ray tubes as the Standard was written for general diagnostic tubes. However, in the absence of other guidance and in the interests of achieving uniform practice it is recommended that the factor of 0.7 is used. Also, note when checking whether equipment complies with DH (2001a) the factor 0.7 is *not* applied.

Equipment: measuring device (slit camera, pinhole or star pattern)

jig or support

mammographic screen–film or non-screen film

magnifying glass (5× to 10×) with graticule (0.1 mm divisions)

Remedial: measured dimension >150% of the nominal value on a reference axis intercepting the breast support platform 5 cm from the chest wall edge for broad focus, or 2 cm for fine focus (for example, >0.65×0.45 mm for 0.3 mm focus, >0.23×0.23 mm for 0.15 mm focus or >0.15×0.15 mm for 0.1 mm focus)

5.6.7 Tube voltage measurement

(a)   Introduction

The most commonly used non-invasive instrument for determining mammographic tube voltage is the digital kV meter. It is important that the measuring device has an appropriate calibration. Correction factors may be required for different target/filter combinations and tube voltage waveforms (Underwood *et al.*, 1996). The digital instrument is easy to use and provides a direct readout of tube voltage for various target/filter combinations and tube voltage waveforms. Suitable exposure factors should be given in the instruction manual. Care should be taken to position the detector in accordance with the manufacturer's recommendations and it may be necessary to angle the detector so that it is perpendicular to the central ray (e.g. by use of a wedge support). With some instruments an oscilloscope may also be used to examine the waveform. This can be useful, particularly if the shape or frequency of the waveform leads to uncertainties in the behaviour of the digital kV meter.

(b)   Range of measurements

Remove the compression plate, as measurements made through the compression plate will be incorrect due to the increased beam hardening. Measurements should be made at a minimum of four voltage settings for the Mo/Mo target/filter combination on both foci and should span the normal clinical range. If a suitable calibration is available, further measurements should be made for other target/filter combinations. If a calibration is not available, it may be useful to perform a limited number of measurements for other target/filter combinations to allow monitoring of constancy.

Suggested exposure factors depend on the measuring instrument (refer to instructions) but are typically 25–50 mAs for broad focus and 20–32 mAs for fine focus. It may be necessary to reduce the focus to instrument distance when using fine focus to achieve the required intensity due to the lower exposure rate. With current calibration procedures, the absolute accuracy of measurement is typically limited to ±1 kV.

Equipment: digital kV meter, oscilloscope to examine the waveform if desired

Remedial: difference between measured tube voltage and set tube voltage >1 kV

Suspension: difference between measured tube voltage and set tube voltage >2 kV

5.6.8 Half value layer and filtration

(a)   Introduction

ICRP (1982) and IPEM (2002) both recommend that for mammography, the total permanent filtration should be equivalent to at least 0.5 mm aluminium or 0.03 mm molybdenum. Details of the filtration will generally be shown on the x-ray tube assembly or given in the accompanying documents. It is required (IPEM, 2002) that every added filter shall be permanently and clearly marked with its filtration in mm of aluminium equivalence (see Section 5.3.3(ii)). The total filtration of a mammographic x-ray tube assembly may be deduced from the half value layer of the x-ray beam measured at a specified tube voltage using published data (see Appendix II). The target angle and tube voltage waveform may also need to be considered.

(b)   Measurement of half value layer

The half value layer (HVL) should be assessed by adding thin aluminium foils to a collimated x-ray beam and measuring the reduction in intensity, ideally in narrow beam geometry. Using a suitable ionisation chamber and electrometer a zero reading without any added aluminium is first established. Further exposure measurements are recorded for each added thickness of aluminium at the same kV and mAs settings.

The aluminium foils must be of adequate purity ($\geq$99.9%) and are attached as close as possible to the tube focus. If the foils are positioned too close to the ionisation chamber, the scatter can result in the half value layer being overestimated. Normally foils with a range of thicknesses of 0.05–0.2 mm to give a *total* thickness up to 0.6 mm are sufficient to define the attenuation curve for the half value layer determination. A more accurate determination of foil thickness can be made by weighing and using the density of aluminium (2.70 g/cm^3) rather than using a micrometer.

Figure 5.3.   Measurement of tube output and half value layer

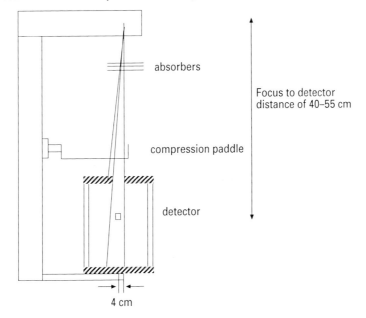

Measurements should be made at 28 kV and additionally for the range of kV and target/filter combinations used clinically. The compression paddle should be present in the beam if the HVL is to be used in the determination of the mean glandular dose (Chapter 7) but care should be taken to avoid the front edge of the paddle. In order to assess the permanent filtration present and to provide a baseline measurement of filtration if, for example, the compression paddle is changed, a measurement should be made without the compression paddle in the beam for each target/filter combination. A jig that collimates the beam to just cover the ionisation chamber and which provides support for the filters can be convenient (see Figure 5.3). This measurement should be made on the same axis as the measurement of output. If the angle with respect to the tube central ray is known, the measurements can be corrected for target angle (Robson *et al.*, 1992).

For a mammography unit with a molybdenum target and molybdenum added filtration, the half value layer will typically be between 0.3 and 0.4 mm of aluminium at 28 kV. This will be greater for other target/filter combinations. Increasing the tube voltage by 1 kV will typically increase the half value layer by 0.01 mm Al. If the measured HVL is 0.4 mm or higher (and if the tube output seems significantly lower than expected) then it may be worthwhile checking whether the light beam mirror inadvertently remains in the beam during exposure. This may affect readings of kV meters as well as HVL measurements.

(c)   Derivation of filtration

The measured HVL (without the compression plate or if a suitable correction is applied to the measurement with the plate) at a known tube voltage can be related to the total filtration using published data (Cranley, 1991; Robson *et al.*, 1992). See also Appendix II.

Equipment:	aluminium foils (≥99.9% purity)
	ionisation chamber and electrometer
	suitable jig or support for foils
Remedial:	HVL <0.3 mm Al at 28 kV Mo/Mo compression plate in
	HVL >0.4 mm Al at 28 kV Mo/Mo compression plate in (note that if other beam conditions are used (higher kV, higher atomic number filter) the measured HVL may be in excess of 0.4 mm Al)
Suspension:	derived total filtration (i.e. compression plate out) <0.5 mm Al or 0.03 mm Mo equivalent
	filtration dislodged, absent, damaged or incorrect (NHSBSP, 1995)

5.6.9   Tube output

(a)   Introduction

It is important to measure the radiation output of the x-ray tube (air kerma/mAs). Too high a value may indicate inadequate filtration of the x-ray beam; too low a value may indicate excessive filtration, target roughening due to prolonged usage or possible problems with the tube voltage waveform. The dose rate is an important factor in determining clinical exposure time.

(b) Measurement of tube output

Tube output should be measured using an ionisation chamber and electrometer system with an appropriate calibration. The chamber should have a near flat energy response over the mammographic range of energies.

The ionisation chamber should be positioned in the x-ray beam on an axis from the focus to a point on the breast support platform 4 cm from the chest wall edge (see Figure 5.3). This position is close to the chest wall edge of the breast support platform in order to avoid variation due to the heel effect but ensures that the chamber is fully covered by the x-ray beam. The chamber should also be at a suitable distance from the tube focus (40–55 cm). If the chamber is of a conventional thin-walled parallel plate or thimble design then it should be positioned at least 10 cm above the breast support platform to avoid possible effects of backscatter. If the chamber inherently provides appreciable backscatter then this distance may be reduced to 5 cm or less. Where the compression plate is included in the measurement, the chamber should not be in direct contact with the compression plate, but in scatter free conditions. The measurements should be expressed as air kerma (μGy/mAs) corrected to a focus-detector distance of 50 cm. For small distance corrections of the order of 10 cm, the effect of attenuation in air can be neglected.

Certain of these tests are performed with the compression plate removed (this eliminates any possible differences in measured output due to, for example, replacement of the compression plate); others are performed with the compression plate in place (for example, determination of data for calculation of breast dose).

(c) Range of measurements

(i) Repeatability of output

Remove the compression plate. Measure the tube output at one setting (usually Mo/Mo, 28 kV) for a tube current-exposure time product of between 20 and 40 mAs. Repeat four times. If problems of erratic exposures have been reported then it may be worthwhile making further checks on output and repeatability at intervals throughout the test period, particularly after the unit has been thoroughly warmed up.

Equipment: ionisation chamber and electrometer

   suitable support

Remedial: maximum deviation of output values from mean >5%

(ii) Specific output

Remove the compression plate. Measure the tube output at 28 kV, broad focus, with the molybdenum target and filter for a current-exposure time product between 25 and 50 mAs.

Equipment: ionisation chamber and electrometer

   suitable support

Remedial: <120 μGy/mAs at 50 cm, 28 kV Mo/Mo

   <70% of output value at commissioning

62 *The Commissioning and Routine Testing of Mammographic X-ray Systems*

(iii) *Specific output rate*

Determine the exposure time for the same current-exposure time product used in (ii) above by direct measurement or infer using known tube current. Using the value obtained in (ii) above for specific radiation output (without the compression plate) and this value for exposure time at the same exposure, the specific output rate can be determined. This should be corrected to a distance equal to the focus to film distance, ignoring the effect of the grid and the table transmission.

Remedial: <7.5 mGy/s at the FFD for 28 kV Mo/Mo, broad focus

Suspension: <5 mGy/s at the FFD for 28 kV Mo/Mo, broad focus

(iv) *Variation of output with tube voltage*

In order to obtain data for the determination of breast dose (Chapter 7) these measurements should be made with the compression plate in place. The entrance air kerma should be measured in a scatter free environment, i.e. at least 5 cm from the compression plate. Measure the tube output at the same values used in the calibration of tube voltage with a tube current-exposure time product of 20–40 mAs. This should be done for broad and fine focus and for each target/filter combination likely to be encountered in clinical use. If desired, the tube output per mAs can be plotted against the kV, which should be near linear (at least between 25 and 30 kV). Using this relationship or by interpolation, the output measurement may be corrected to a standard kV, thus allowing for drift in the tube voltage (e.g. to determine effect of tube age).

Equipment: ionisation chamber and electrometer

            suitable support

Remedial: none

(v) *Variation of output with mAs*

Remove the compression plate. Measure the output at a single selected value of tube voltage (usually 28 kV with a molybdenum target and filter or the combination most frequently selected in clinical use) for a range of mAs values. The range should encompass the clinical range for both broad (typically 10–400 mAs) and fine focus (typically 10–150 mAs). It may be appropriate to test other target/filter combinations if the tube current is significantly different from that used for the molybdenum target. Any variations may be due to problems in tube current calibration (assuming the tube voltage and exposure time to be correct).

Equipment: ionisation chamber and electrometer

            suitable support

Remedial: maximum deviation of output/mAs from mean >10%

5.6.10 Anti-scatter grid

(a) Introduction

Most mammographic examinations are performed using an anti-scatter grid. A focused grid is mounted in the Bucky assembly and moves during the exposure. In addition to linear grids, recent developments include a cellular grid that rejects scatter in two directions (see Section 3.7).

The grid ratio, line density and focal length will normally be shown on the grid itself or given in the accompanying documentation. For a moving grid, DH (2001a) recommends a grid ratio of 4:1 or 5:1 with a line density of approximately 30 lines/cm. The method of assessing the performance of anti-scatter grids described in British Standard 61953 (BS, 1998) is applicable to mammographic grids, but tests are not intended for field measurements. Additional information has been given by Keevil *et al.* (1987).

The *grid factor* gives the increase in exposure required when using the grid, and in mammography is typically about 2 and should not be greater than 3 (DH, 2001a). It is defined as the ratio of the air kerma incident on the cassette with the grid in place to the incident air kerma without the grid and is therefore a property of the grid itself and not of the grid system. The grid factor is easily measured in systems where the grid alone can be removed, but in the many cases where this is not possible, the *grid system factor* may be measured instead. The grid system factor is defined here as the ratio of the incident air kerma with the grid system in place to the incident air kerma without the grid system. As an approximation, the ratio of mAs values may be used in place of the ratio of air kerma. For a Bucky (moving grid) system, the grid factor and the grid system factor will differ because of absorption by the breast support platform. If a 24×30 cm Bucky assembly is provided and is used clinically, the tests described should be repeated.

(b) Measurements

(i) Grid factor/grid system factor

The grid factor should be determined using a molybdenum target and filter at 28 kV with a 4 cm Perspex phantom (to produce scatter) on the breast support platform above the plane of the grid. A film method employing a mammography cassette can be used. The grid factor is estimated from the ratio of the tube current exposure time product (mAs) required to produce the same optical density on the processed film with and without the grid. The grid system factor may be estimated in a similar way with the cassette being placed on top of the cassette tunnel for the exposure without the grid and a small inverse square law correction being applied to correct for focus-film distance.

$$\text{GSF} = [\text{mAs}(1)/\text{mAs}(2)] \cdot [\text{FFD}(2)/\text{FFD}(1)]^2 \tag{5.4}$$

(ii) Grid condition and artefacts

It is useful to produce an image of the grid at the focus–film distance at which it is used. This allows the line density to be estimated (through a magnifying lens) and the uniformity of air kerma rate across the radiation field to be assessed. Poor uniformity may be due to grid cut-off that indicates a misaligned grid or a focused grid of the incorrect focal length.

Figure 5.4. Measurement of grid system factor

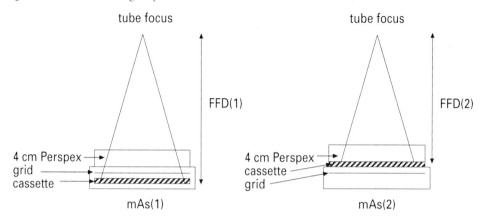

The grid movement should ideally be disabled for this examination, but if that is not possible it may be sufficient to make a very short exposure using AEC with no additional attenuation in the beam. Such an exposure will usually show up any non-uniformity in the grid and breast support platform cover and may also show the grid lines and enable the line density to be estimated.

Equipment: 4 cm Perspex phantom

   densitometer

Remedial: line density outside manufacturer's specification

   grid factor >3.0 (DH, 2001a)

5.7 Mammography system tests

5.7.1 Introduction

In general, the tests described in this section are of the whole imaging system comprising the x-ray unit, screen–film combination and processor. In many of these tests the x-ray unit cannot be isolated from the rest of the system as, in addition to the choice of x-ray exposure parameters, the processing conditions and choice of screen–film combination can significantly affect image quality and breast dose. This is further discussed in Chapters 6, 7 and 8.

5.7.2 X-ray image uniformity

(a)   Introduction

It is important to assess the image uniformity and to identify any artefacts for several reasons.

- Non-uniformities in optical density between the left and right sides of the x-ray field may be significant when viewing mammograms.

- A mammographic x-ray tube is normally designed such that the part of the x-ray beam having the highest air kerma rate is directed towards the thickest part of the breast, i.e. closest to the chest wall. If this is not the case, the tilt angle of the x-ray tube in the housing may be incorrect.

- The thin foils used as beam filters may not be of a uniform thickness or may have perforations or imperfections.

- Artefacts may be superimposed on a clinical film by deposits on the x-ray tube window or on the beam filters.

- Problems with grid uniformity or grid movement may be highlighted.

- There may be problems with the uniformity of the intensifying screens in the cassettes.

The test method described below employs film and is similar to the British Standard for the determination of the maximum symmetrical radiation field (BSI, 1985), although this standard is not applicable to mammographic x-ray tubes. However, no attempt is made to calibrate the film density in terms of air kerma or to position the cassette and aluminium sheet perpendicular to the reference axis of the tube. Perspex slabs can be used instead of aluminium, although the results may differ slightly. Care should be taken to ensure that the exposure is sufficiently long to prevent the appearance of grid lines on the image. If a 24×30 cm Bucky assembly is provided and is used clinically, the tests described should be repeated with this assembly in place.

Procedures for testing the uniformity of small-field digital systems can be complex and system specific and are fully described in NHSBSP Report 01/09 (2001).

(b)   Method

Place a sheet of aluminium, 2–3 mm in thickness, or Perspex slabs, approximately 4 cm in thickness, on the breast support platform. The material should completely cover the imaging area of the cassette. Alternatively, use a smaller sheet of aluminium (typically 100×100 mm) at the tube port to give complete beam coverage. If using aluminium sheet ensure that the sheet is flat and free from major flaws. With a loaded cassette in the Bucky, make an exposure at 28 kV to give an optical density on the processed film of between 1.5 and 1.9 (on the midline at 4 cm from the chest wall edge). The compression plate should be removed in order to exclude possible variations due to its structure (the test may be repeated with the compression plate in place to identify any non-uniformity due to its structure).

Measure the optical density in the centre of the field (A) and 10 cm either side of the midline (B), 4 cm from the chest wall (see Figure 5.5). It may be desirable to measure densities at 2 cm intervals across the film, if large deviations from the midline are found. Inspect the film for any local variations in optical density, caused by defects in the construction of the x-ray unit (e.g. which might indicate imperfections in the filter or grid or foreign matter in the tube head). Processing wash marks on the film may also cause large deviations. Repeat for all other target/filter combinations and fine focus

Figure 5.5. Positions for measurement of optical density for uniformity assessment

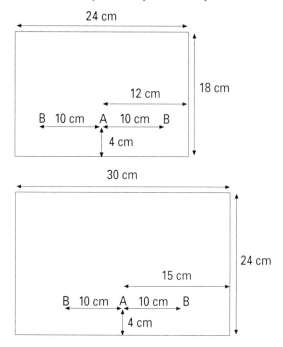

(magnification mode) where appropriate. It may also be appropriate to repeat the measurement with the cassette placed on the breast support platform (under the Perspex or aluminium). This will give information about whether any non-uniformity is due to the grid or breast support platform or is in the x-ray beam. If it is suspected that the intensifying screen in the cassette is causing the non-uniformity, the measurements can be repeated using non-screen film or a different cassette. Typically, a 20% decrease in optical density from chest wall to nipple edge is observed for an 18×24 cm film using broad focus. This will be greater for fine focus or 24×30 cm images.

Equipment: aluminium sheet, 2–3 mm in thickness, or Perspex slabs, 4 cm in thickness (18×24 cm or 24×30 cm) having a uniformity of thickness of better than ±0.5%

screen–film cassette or envelope wrapped non-screen film

densitometer

Remedial: difference in optical density points at centre of field and 10 cm either side of centre (A-B) >0.2

difference in optical density between left and right points 10 cm either side of centre >0.15

the cause of any significant artefact(s) should be investigated

5.7.3 Automatic exposure control system

(a) Introduction

On mammography x-ray units designed for screen–film mammography, a radiation detector located after the image receptor normally controls the duration of the radiation exposure. The detector monitors the x-rays transmitted by the receptor and the exposure is terminated when the radiation dose received by the detector reaches a predetermined level corresponding to the desired optical density on the film.

The position of the radiation detector can be varied between two or more predetermined positions to facilitate the exposure of breasts of different size. A density control can be used to vary the exposure above or below the predetermined level and a further control adjusts for different screen–film combinations.

In the latest generation of equipment, the unit may also have a method of optimising the beam quality (target/filter combination and tube voltage) for the individual breast. The choice is dependent on the thickness of breast tissue under compression or a combination of this and the composition of the breast.

A guard timer is fitted which prevents gross over-exposure of the breast if the automatic exposure control system fails. The guard time may be related to the maximum mAs available or to the manual setting of mAs. In the latter case it is important to ensure that the value selected does not result in premature termination of the exposure. Certain machines are capable of monitoring the dose rate received by the AEC detector and may terminate the exposure after a short duration if the dose rate is so low that insufficient mAs will be available to complete the exposure.

The performance of the automatic exposure control system can be described in terms of three factors:

- Repeatability of the system for exposure under identical conditions.

- Variation of the response of the system with radiation quality (there may be significant changes in radiation quality with breast thickness and composition due to changes in the x-ray spectrum of the radiation transmitted through the breast and the use of different target/filter/tube voltage combinations).

- Variation of the response of the AEC system with exposure rate.

For reliable and consistent operation, it is important that the energy responses of the detector and screen–film system in use are well matched. Also, the AEC will need to cope with a range of exposure rate of up to two orders of magnitude.

The automatic exposure control should be tested using a phantom that simulates breasts of different compressed thickness. The phantom normally consists of a series of Perspex slabs, each 1 or 2 cm in thickness, and of semicircular (at least 16 cm in diameter) or rectangular cross-section. Although the shape and area of the horizontal cross-section through the phantom can affect the performance of the AEC, using the same slabs each time the unit is tested will ensure constancy. To allow for possible small variations in the thickness of the Perspex slabs, the individual slabs should be numbered so that the same combination of slabs can be used in the same orientation each time the tests are performed.

Breast composition changes with breast thickness (Chapter 7), so it may be more appropriate to perform some tests with a better tissue equivalent material than Perspex.

This is particularly relevant when testing systems with automatic beam quality systems that can select the kV, filter and target material. However, this is not currently routine practice.

For the following tests the screen–film combination normally used for mammography should be employed. The same cassette should be used for all exposures with the films taken from the same batch. The same processor should be used and appropriate data on the processing parameters (e.g. chemistry type, developer temperature, cycle time, replenishment rates) should be recorded. If a 24×30 cm Bucky assembly is provided and employs a different AEC detector or if any other AEC detector is used, the tests described should be repeated.

Note that some of the AEC tests are not categorically system tests but are included in this section for convenience.

(b) Test procedures

For all tests of the automatic exposure control system, the optical density of the processed films should be measured at a standard position on the midline of the film, 4 cm from the chest wall edge of the film (see Figure 5.5 (A)) (to match the position of tube output measurement, Section 5.6.9).

(i) Target density

The automatic exposure control should be set up to give the desired optical density (the target density) on the processed film. A value between 1.5 and 1.9 is recommended (NHSBSP, 2004). This is normally set by the service engineer on installation and must be correct for the screen–film system and processor in use. A 4 cm thickness of Perspex should be used, positioned as for a cranio-caudal projection with the slabs slightly overlapping the chest wall edge of the breast support platform. The AEC detector should be positioned at the chest wall edge with the compression plate in place against the top surface of the Perspex. A standard compression force or compressed thickness should be used as, depending on the design and operation of the AEC system, these may affect the selection of exposure parameters. The x-ray unit should be operated in the usual mode with the density correction and screen–film selection as for normal clinical practice. For systems without beam optimisation software, a tube voltage of 28 kV with a molybdenum target and molybdenum filter or typical clinical exposure factors is suggested. Where a magnification facility is provided, measurements should be repeated in this mode (this will normally use fine focus).

Remedial: maximum deviation in optical density for 4 cm thickness of Perspex from target density >0.2 or outside the range 1.5–1.9

Suspension: optical density (for 4 cm thickness of Perspex) outside the range 1.3–2.1 and not correctable by adjustment of AEC density control

If the measured optical density is outside the remedial level it may be desirable to adjust the density control before the following tests are performed. Investigations should be made as to the cause of the difference between the measured optical density and the target density.

(ii)   *Overall repeatability*

Make at least four exposures with 4 cm thickness of Perspex in the beam and record the mAs for each exposure. A loaded cassette should remain in place for all the exposures. In addition, make an exposure in each AEC detector position (moving the slabs to cover the detector area if necessary) and record the mAs for each exposure. If the mAs readings vary by more than 10%, one of the detector positions may be obstructed. Some systems may have multiple detectors rather than a single moving detector. If these can be selected independently or in different combinations, it should be verified that all the detectors or combinations have a similar response.

If a problem is suspected with indicated post-exposure mAs, an ionisation chamber may be positioned in the beam so that it does not shadow the AEC detector, and used to check constancy of operation. If investigating reports of inconsistent behaviour for particular clinical views, it may be found valuable to repeat the tests with the x-ray tube and breast support platform angled to the two extreme lateral positions to check for loose connections, etc.

Remedial: maximum deviation of mAs from mean >5%

Suspension: maximum deviation of mAs from mean >10%

(iii)   *Constancy with change in thickness*

Use 2, 4, 6 and 7 cm thicknesses of Perspex to simulate breasts of different size. On broad focus, using the mode of operation and settings used clinically, make one exposure for each thickness. Repeat for fine focus (magnification mode) if this is used clinically. Record the exposure factors (target, filter, kV and mAs). If automatic beam quality optimisation is not available or selected, the measurements should be made using the tube voltage that is used clinically. If, for larger thicknesses of Perspex, the guard or back-up time is reached the kV will need to be increased.

For machines with automatic beam quality optimisation, it may be necessary to repeat the measurements using a single target/filter/tube voltage combination for each thickness, in order to determine whether the observed variations are due to change in breast thickness or beam quality. Additional measurements may be needed using other thicknesses of Perspex depending on the design and mode of operation of the AEC system.

Remedial: maximum deviation in optical density from value with 4 cm Perspex >0.20 or range of optical densities >0.30 or optical density outside the range 1.3–2.1

Suspension: maximum deviation in optical density from value with 4 cm Perspex >0.40 or range of optical densities >0.60

(iv)   *Constancy with change in tube voltage*

For systems without automatic beam quality optimisation use 4 cm Perspex. Make exposures for at least four tube voltage settings, which cover the range used clinically.

Remedial: maximum deviation in optical density from value with 4 cm Perspex >0.20 or range of optical densities >0.30 or optical density outside the range 1.3–2.1

Suspension: maximum deviation in optical density from value with 4 cm Perspex >0.40 or range of optical densities >0.60

(v)   *Constancy with change in other operating parameters*

Make a series of exposures that will test the effect of any other variable function on the operation of the AEC system (e.g. change of filter or target).

Remedial: maximum deviation in optical density from value with 4 cm Perspex >0.20 or range of optical densities >0.30 or optical density outside the range 1.3–2.1

Suspension: maximum deviation in optical density from value with 4 cm Perspex >0.40 or range of optical densities >0.60

(vi)   *Density control*

It is useful to test the calibration of this control to provide an indication of the density change per step. The conditions described above for measuring overall constancy can be used. The mAs and density reading should be recorded for appropriate positions of the density control and the results expressed as a percentage of the mAs or change in the optical density from the normal position. The steps should be small enough to enable useful adjustments to be made in clinical use, especially where high contrast screen–film combinations are used. Current machines should be capable of 5–10% steps in mAs (DH, 2001a), in some cases the step size can be altered by the service engineer.

(vii)   *Guard timer*

Before commencing this measurement, it is important to refer to the tube rating data to ensure that any exposure made is within the rating of the tube. Position approximately 8 cm of Perspex or a suitable thickness of an alternative attenuating material (e.g. 0.5 mm Pb sheet) on the breast support platform so that the AEC detector is covered. Make an exposure and record the mAs. If necessary increase the thickness of attenuator until the guard time is reached (this is usually indicated by an error message and/or audible warning). Allow an adequate time between exposures to prevent overheating the x-ray tube. On some systems the AEC detects very low dose rates and will terminate the exposure immediately (again an error message and/or audible warning is usually given).

The nominal guard time should be ascertained before any exposures are made, so that the exposure can be terminated manually should the cutout fail. If any exposure lasts long enough to be terminated by the guard timer or to require manual termination, *wait at least 5 minutes* before any further exposure is made or refer to the system rating charts regarding the required waiting time.

(viii)   *Exposure time*

Most mammography x-ray units control and indicate exposure duration in terms of mAs rather than exposure time. However, it may be useful to determine the exposure time for typical clinical exposures. Excessive exposure times can result in the increased likelihood of movement artefacts. Changes in exposure time may indicate problems with tube current calibration. The exposure time can be approximated from the mAs and nominal tube current (this does not take into consideration changes in calibration of mA) or measured using a digital exposure timer or waveform from a suitable detector displayed on an oscilloscope. Note that on some mammography x-ray systems, at longer exposure times the exposure may momentarily stop as the grid movement changes direction (to prevent the appearance of grid lines on the film). This may affect the reading on certain designs of digital exposure timer.

Remedial: exposure time using typical clinical parameters for 4 cm of Perspex >1 s

exposure time using typical clinical parameters for 7 cm of Perspex >4 s

(c)   Equipment for testing automatic exposure control systems

- Numbered slabs of Perspex with thicknesses in the range 2–8 cm.

- Ionisation chamber and electrometer (if necessary).

- Densitometer.

5.7.4   Automatic exposure control for small-field digital mammography systems

(a)   Introduction

For small-field digital mammography devices, the AEC system may use the image receptor to monitor the radiation dose and terminate the exposure. The tests for the automatic exposure control are similar to those for screen–film systems (overall repeatability and variation with change in Perspex thickness). The Perspex attenuator should be positioned to fully cover the image receptor. Whereas deviation in optical density is measured for screen–film, the average pixel value at the centre of the image, either using region of interest analysis or choosing individual pixel points, and the post-exposure mAs should be recorded for small-field digital systems. The repeatability tolerances for AEC performance have been taken from NHSBSP report 01/09 (2001) 'Commissioning and Routine Testing of Small-field Digital Mammography Systems'. Although no accepted guidance on the optimal exposure factors for digital systems is available, it is unlikely to be the same as for screen–film systems and may well differ from one type of digital system to another. Thus, although it is important to test and record how each AEC system is working, a specific remedial level when testing with different thicknesses of Perspex is not provided and users should consult the manufacturer for guidance.

(b)   Test procedures

(i)   *Overall repeatability*

Using 4 cm Perspex, measure the average pixel value and record the mAs.

Remedial: commissioning – maximum deviation of mAs and average pixel value from series mean >5% (baseline values set at commissioning)

routine – maximum deviation of mAs from baseline value >5%, maximum deviation of average pixel value from baseline value >10%

(ii) *Variation with change in Perspex thickness*

Using 2, 4, 6 and 7 cm Perspex, measure the average pixel value and record the exposure factors.

Remedial: consult manufacturer's guidance

CHAPTER 6

The testing of processors and screen–film systems

6.1 Introduction

The tests described in this chapter relate to the commissioning and subsequent periodic routine testing of automatic processing units (APUs), films, screens and cassettes, the assessment of darkroom light safety, viewing conditions and film processing artefacts. It is anticipated that most of these tests will normally be undertaken by radiographic staff or manufacturers' service engineers, although periodic testing of APU performance and the monitoring of image artefacts also form part of the range of tests undertaken by physics staff. This is a valuable addition to the more frequent routine tests undertaken by local mammography personnel because it provides an external reference, as any deterioration in system performance can be gradual and may not be apparent from daily routine checks. It is important that physics staff are involved in the commissioning of film processing systems, screens and cassettes to enable system optimisation. Physicists should be available to provide advice and support to those conducting routine tests and must be are aware of the significance of the tests, how they are undertaken and how to interpret the results. They should be familiar with the QA manual for Radiographers produced by the College of Radiographers, 'A Radiographic Quality Control Manual for Mammography' (NHSBSP, 1999b).

A comprehensive account of the tests undertaken in the radiology department on APUs, films, intensifying screens, cassettes, illuminators and the darkroom is provided in IPEM (1996). In this chapter, the differences between the general and mammography situations are highlighted and the emphasis placed on how the tests should be undertaken and the results interpreted.

The ultimate quality and diagnostic value of the mammographic image depends critically upon a number of factors, including the type and condition of the film, intensifying screens and cassette and the processing of the latent image after exposure. In mammography, this is at least as important as the performance of the x-ray set in determining overall image quality. A significant proportion of problems identified with mammographic image quality arise as a consequence of problems with the screen–film system or the APU. It is therefore important that the appropriate screen–film combination is used and procedures undertaken to optimise performance of APUs at the outset. The optimisation process should be repeated whenever major changes such as the introduction of a new film type or processing chemicals take place. Having identified the correct operation and adjustment of the APU and associated equipment, the level of performance achieved should be carefully monitored using the procedures described in this chapter.

6.2 Background to monitoring film processor performance

Film sensitometry provides an objective measure of film characteristics that directly reflect both the processing conditions and inherent characteristics of the film. Three parameters, base plus fog, photographic speed and film contrast, are traditionally used to define the shape of the sensitometric curve. Changes in these parameters reflect variations in the processor or film that may significantly affect image quality.

Stepwedge films may be produced by light or x-ray sensitometry. For practical reasons, a light sensitometer is the instrument of choice. It consists of a highly reproducible light source and reference stepwedge. The light output of the sensitometer is usually adjustable in several increments of 0.15 log exposure (log E) to suit the photographic speed of the film under test. The correct level is determined at commissioning of the system so that the full sensitometric curve is demonstrated. The emitted light spectrum is often adjustable between 'green' and 'blue', with the 'green' setting being suitable for most current mammography films. The film response curve produced (Figure 6.1) relates to the light from the sensitometer, which does not precisely match the situation when the film is exposed by light produced from an intensifying screen. The spectral output of the intensifying screen is usually different to that of the light sensitometer and the overall exposure time may lie outside the range encountered clinically. This may cause a further difference in results arising from reciprocity failure effects.

Figure 6.1. The sensitometric curve and sensitometry indices

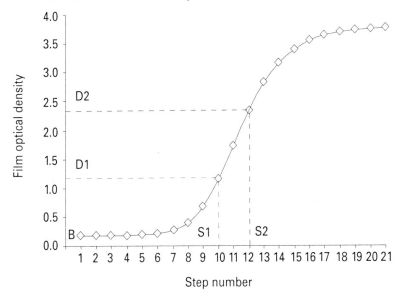

Experience has shown that light sensitometry is a sufficiently good surrogate for x-ray sensitometry and is usually adequate for the purposes of APU optimisation and routine monitoring (West and Spelic, 2000). It offers a significant practical advantage to the user over x-ray sensitometry in that the exposures generated are reproducible within a tight tolerance and reliance on the exposure reproducibility of an x-ray system is avoided.

Variation in sensitometric indices can arise as a consequence of orientation of the film and film emulsion in the APU. The user should always feed the film with the sensitometric strip on the leading edge and with the manufacturer's recommended orientation of the emulsion.

Another important variable is that of latent image fading. This arises when there is a delay between irradiation of the film and subsequent processing. The magnitude of the effect varies with film type and is illustrated in Figure 6.2 for one film type. Ideally, the film should be processed immediately after exposure or if this is not possible, a fixed period of time allowed between irradiation and processing. The use of pre-exposed strips for monitoring APU performance is not recommended.

Figure 6.2. Effect of delay between exposure and processing of film upon film speed and optical density of a film taken of a 4 cm Perspex block

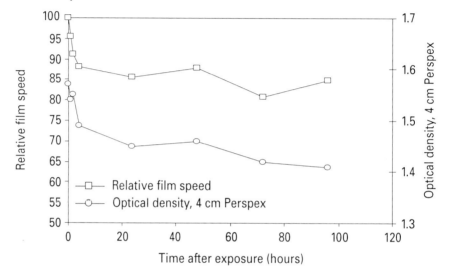

6.2.1 Sensitometric parameters

Three principal sensitometric parameters are illustrated in Figure 6.1.

(a) *Film base plus fog*. The optical density, B, of an unexposed region of the film or the lowest optical density point on the stepwedge.

(b) *Film photographic speed*. The exposure required to produce a given film optical density. For mammography, it is recommended that an optical density of 1.5 above base plus fog be used in order to reflect clinical practice. The relative light exposure required to produce a given film optical density, can be calculated from optical density measurements of the stepwedge film, using interpolation or curve fitting algorithms. This assumes a fixed and known increment in exposure between adjacent points on the stepwedge. Film speed can be normalised and expressed in relation to the speed of a standard film, or the film speed calculated can be used as an arbitrary quantity solely for the purposes of quality control. This is the approach adopted by most PC based sensitometry packages. Alternatively, a simple speed index parameter, S, suitable for manual sensitometry systems can be defined as the step chosen with a density D_1 approximately 1.0–1.5 above B. $S = D_1 - B$ (or alternatively D_1 alone), can be shown to be sensitive to changes in film speed and is used as a surrogate of film photographic speed.

(c) *Film average gamma or gradient.* This is the average gradient between two defined points of the characteristic curve. Several parameters are in common use, including average gradient between optical densities 0.25 and 2.0 above base plus fog ($G_{0.25-2.0}$), a broad range measure of film contrast, or average gradient between optical densities 1.0 and 2.0 ($G_{1.0-2.0}$), a midrange measure of film contrast. This is the approach adopted by most PC based sensitometry packages. Alternatively, a simple contrast index suitable for manual sensitometry systems can be defined as $C = D_2 - D_1$, where D_2 is a step chosen with a density approximately 2.0–2.5 above B.

In general, the sensitometric speed and contrast parameters generated by PC based sensitometry packages are to be preferred to corresponding indices used for manual sensitometry systems. This is because the parameters calculated by the former are independent of each other whilst those calculated by the latter are inter-dependent. This could lead to ambiguity in the interpretation of results.

When undertaking sensitometry on a daily basis, film base plus fog, speed and contrast indices are plotted as a function of time (Figure 6.3). When the parameters are so displayed, trends or slow variation in the parameters tend to be more apparent and can then be acted upon before changes become significant in terms of the quality of the radiographic image. Further measurements may then be required to identify the cause of these variations and some are described in Sections 6.3.2 and 6.3.3.

6.2.2 Applications of film processor sensitometry

(a)   Initial commissioning and optimisation

The installation engineer will normally adjust the APU to a standard set of parameters. However, the optimum set of parameters will usually depend upon a range of factors including workload, APU type, processing chemicals and range and type of films processed in addition to the preferences of the radiologists or film readers. Further adjustments to the APU may be required subsequent to installation in order to achieve the optimal situation.

The first stage in optimising the APU is to consider how changes in developer temperature, processing cycle time or other relevant variables affect film base plus fog, speed and contrast. Using the method of light sensitometry, the values of these parameters should be determined for the likely range of variables including, for example, developer temperature and cycle time. In each case, sufficient time should be allowed for the developer tank to reach a stable temperature.

Broadly speaking, for mammography the processing regime is selected which maximises film contrast and speed within the constraint of acceptable film base plus fog level. However, there may be other factors to take into account such as occurrence of film emulsion 'pick-off' (which has the appearance of small white 'specks' on the processed film) or roller marking (which can take a variety of forms), which tend to increase with developer temperature. Extent of aerial oxidation of the developer solution also tends to increase with temperature and may lead to instability of the processing system with changes in workload or when the system is left unused.

76 The Commissioning and Routine Testing of Mammographic X-ray Systems

Figure 6.3. Example of control charts suitable for displaying the results of sensitometry measurements

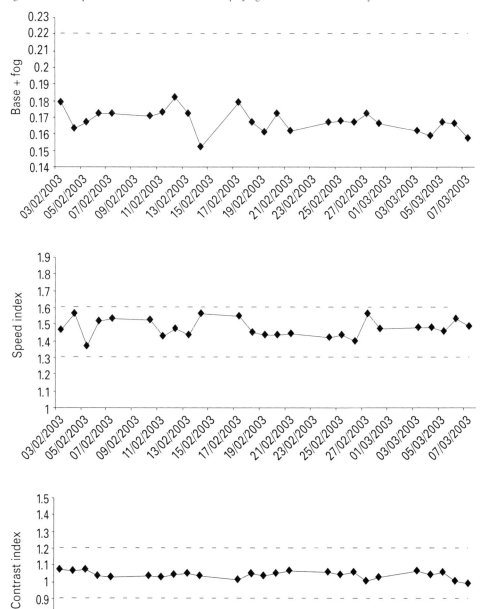

Choice of developer and fixer replenishment rate can also have a significant effect upon APU performance. The relationship between this and the sensitometric parameters can be difficult to predict and is often specific to the type of film and processing chemicals used. Additionally, the use of a silver recovery or fixer re-circulation system will influence optimum replenishment rate. Usually, the lowest replenishment rates are selected that are consistent with the optimum level of performance.

Whenever changes are made to the processing parameters, it may take some time for the full effect upon the processing system to become apparent. As a 'rule of thumb', the developer tank needs to be replenished with a volume of fresh developer equal to three times its capacity before equilibrium is attained. Depending upon the developer tank size and the workload, this may take between a few days and several weeks to occur. Therefore, once an APU has been adjusted, it may be necessary to re-examine the system after equilibrium has been attained to ensure that optimum parameters have been selected.

Where more than one film processor is in use, it is desirable to match their performance so that they may be used interchangeably. Care should be taken that an optimum level of performance for all processors is achieved and numbers of films processed equalised between the systems.

(b)   Establishment of baselines

APU sensitometric data should be collected for at least three to five days or until an apparent equilibrium position is established. If a PC based sensitometry package is used, an automatic routine for establishing baselines will usually be provided which allows the user to set limits for each parameter. These limits can then be reset periodically to reflect changes in APU performance arising from seasonal variations or changes in workload or practice. It is important that whenever baselines are altered, care is taken that the APU remains optimised and it may be appropriate to adjust APU parameters in order to regain optimum performance. Reference should be made to the manufacturer's specification and the results of any national or regional reviews.

(c)   Routine monitoring

Routine monitoring should be carried out every day the APU is used, or more frequently if problems are encountered. Monitoring using a film exposed with a sensitometric stepwedge on only one side is sufficient. The tolerance bands within which the sensitometric parameters are allowed to vary should reflect the allowable variation in APU performance and the systematic uncertainties implicit in the measurement process. If operating a manual system, measurements should be plotted on a chart in addition to being noted numerically as the graphical display of parameters reveals fluctuations more effectively than inspection of numerical data (Moores *et al.*, 1987).

Usually, the best time of day to conduct these tests is after the processing system has been switched on and allowed to stabilise and before clinical images are processed. However, results recorded from sensitometric strips processed at the beginning of the working day may not be typical of results achieved later in the day. The users should therefore be aware of how sensitometric parameters vary throughout the working day, if necessary by establishing this experimentally.

6.2.3 Causes of variation in sensitometric parameters

Variation in sensitometric parameters can be due to the film. This source of variation can be minimised by using the same box of film to produce the sensitometric strips, although this can lead to problems with gradual fogging due to storage of the film. An alternative approach is to use film from the main stock. In this case, film sensitometric indices will reflect those achieved by the film in clinical use.

Variations in base plus fog level will normally be within about ±0.02 over periods of several weeks or months. Variations outside this range may, for example, be due to the condition of the chemicals in the processor, changes in film emulsion numbers or batch numbers, changes in developer temperature, darkroom light leaks and inadequate safelights, or film storage. If conditions in the darkroom are sub-optimal, this will tend to manifest itself as variations in film contrast and speed before base plus fog is affected. Variations in film contrast are often smaller than those in film speed but both should normally be less than about 5%.

Film speed is usually the most variable of the three quantities. Although it may remain within a tight tolerance for short periods, it is uncommon for this to be maintained over many months. Changes in the film speed index greater than the allowed range of tolerance can occur, even from one day to the next. Sudden changes can often be traced either to a change of film emulsion or batch number, or to the processor being serviced with fresh chemicals. Little can be done about the first; the speed index may return to its former level with the next change of film number. This cause may be confirmed if a formal routine is adopted of checking variation on sensitometric parameters with change of film batch number.

Before investigating variations, the random variation to be expected from the sensitometer and densitometer must be known approximately, although this should be apparent by observing the range of daily random variations in speed and contrast.

When processors are serviced, the processor is often refilled with new chemicals. In order to bring the developer solution swiftly back to its equilibrium state, special 'starter' solutions may be added to the fresh developer. Servicing engineers may vary the proportions according to their estimate of local needs, or to instructions given by manufacturers. A sharp drop in film speed or speed index is often observed on such occasions and may take a while to regain its former level. In practice, the approach has been adopted where part or all of the 'seasoned' developer solution is retained whenever the processor is emptied. This may not eliminate the problem but will usually reduce it and may allow a more rapid return to former values of speed index.

6.3 The automatic processing unit

The checks on the APU undertaken by physics staff are described here and the frequency at which they should be undertaken summarised in Appendix I. They are usually undertaken in order to:

- Monitor changes since the previous physics survey and make comparisons with standard data.

- Assist in interpretation of changes in image quality and dose.

- Identify long term trends, in conjunction with inspections of local daily sensitometry measurements.

Additional checks on specific gravity, silver content of fixer, residual fixer on film and replenishment rates can be useful fault finding tools, but are normally performed by processor engineers. Additional information can be found in the manufacturer's literature (IPEM, 1996; NHSBSP, 1999b).

6.3.1 Light sensitometry

(a) Method

A stepwedge film is produced by exposing a sheet of film under darkroom conditions using a light sensitometer, taking care to expose the emulsion side of the film. Ideally, the stepwedge should be exposed onto the leading edge of the film with respect to the orientation in which it enters the APU. If this is not possible, the stepwedge should be exposed such that the lower optical density end of the stepwedge is fed into the developer tank first (BSI, 1994a). The optical densities on the processed film should be measured, and, as a minimum, values of film base plus fog, speed and contrast calculated as described in Section 6.2.1.

(b) Equipment

Light sensitometer

Densitometer

PC based sensitometry package (if available)

(c) Results

Comparison should be made with previous measurements and professional judgement applied in order to determine whether parameters have drifted from optimum values. Examination of the data for the presence of longer-term trends should be undertaken periodically, either as part of physics staff routine tests, or at appropriate intervals, in conjunction with local radiography staff.

6.3.2 Temperature

The temperature of the developer tank should be measured directly with an appropriate calibrated alcohol–glass thermometer or digital thermocouple. A glass mercury thermometer should not be used because of the danger of contamination if it gets broken. Measurements should be taken at several positions around the developer tank and a mean value calculated to determine the accuracy of the processor's temperature display. It may also be necessary to repeat the measurements in order to determine the stability of the heating circuits.

Remedial :   stability >0.5°C or outside manufacturer's specification

accuracy >0.5°C or outside manufacturer's specification

6.3.3   Transport speed

The total transit time of a film through an APU can be determined using a stopwatch. The time interval measured should be from entry of the film's leading edge to its first reappearance, or entry of the trailing edge to complete emergence of the processed film. With a daylight processing system, it may be difficult to determine the point at which the film enters the developer tank. In this case, attention should be paid to the sounds made by the film processor, characteristic of when the film enters the developer tank.

Remedial: difference >5% of manufacturer's specification

6.4   Calibration of test equipment

6.4.1   Calibration of densitometers

Most spot densitometers and automatic strip readers are supplied with a calibrated reference strip having three or more reference optical density areas. The optical density of each area is checked under standard measurement conditions.

It should be noted that the accuracy of reference strips is only guaranteed for one to two years from the date of purchase. Most stand-alone densitometers can be calibrated by the user and reference should be made to the manufacturer's recommendations. Densitometers used in conjunction with computer based sensitometry packages can also be calibrated, usually by scanning the reference strip in calibration mode and entering the optical densities of the reference points into the software.

Remedial :   difference >0.02 of value specified at reference points, up to a density of 3.0 or outside manufacturer's specification

Suspension: difference >0.05 of value specified at reference points

6.4.2   Inter-comparison of sensitometers

(a)   Introduction

The purpose of this test is to establish the performance of the sensitometer against a standard instrument and to establish stability of the unit over a period of time. Absolute calibration of the light output of a sensitometer is difficult to achieve and the method of inter-comparison is recommended.

Light sensitometers are known to vary with time, and even between nominally identical units. Any inconsistency in the exposure increments will lead to differences in sensitometric indices measured by different instruments, as will any variation in the output or uniformity of the light source or deterioration of the stepwedge over time.

Variations in sensitometry occur even between instruments of the same model with sequential serial numbers with differences in calculated contrast varying by up to 25%.

The primary function of the sensitometer is to enable tests of constancy with time of a given combination of a single processor and a single type of film. If sensitometers are used for other functions, such as establishment of absolute speed and film gradient parameters, care should be taken to calibrate the instrument. Where absolute measurements are required, x-ray sensitometry is usually the method of choice.

(b)   Method

At least three sensitometric strips should be exposed with both the standard sensitometer and the local instrument under identical conditions. Allow the sensitometer to rest for 30 seconds between exposures in order to obtain consistent light output from the device. The films should be processed sequentially through the same APU. Values of speed and contrast parameters are calculated for the films exposed by both sensitometers.

(c)   Equipment

Standard or reference sensitometer

Local sensitometer

(d)   Results

Remedial: relative values of speed and contrast should not change by more than 5% from the baseline measurement

6.5   The cassette–screen–film system

Tests of new screens and cassettes are important to ensure comparability within a batch, or with existing cassettes. Any variation between cassettes introduces a further source of variation in optical density of the mammograms which may be comparable to or greater than that encountered with the normal variation in film processor sensitometry. Routine tests at suitable intervals are also required to detect any deterioration; cassettes may deteriorate mechanically, and screens may lose sensitivity and speed at variable rates or be damaged. Screen–film contact is particularly important, and must be checked when cassettes are new and regularly thereafter. Further guidance may be found in DH (1993). These tests are normally undertaken by radiographers, although may also be undertaken by physics staff as part of an investigation.

6.5.1   Screen–film contact

(a)   Method

Load all cassettes to be tested with fresh film and allow them to 'dwell' for a period of ten minutes. Lay the test tool flat in contact with the top of a cassette and expose to produce a film density of 1.5–2.0. (Refer to manufacturer's instructions for further details.) The cassette should be positioned on top of or in place of the cassette holder and not in its normal position beneath the grid. Take care to position the cassette carefully so that the film is covered by the x-ray field, particularly at the chest wall edge. Repeat for all cassettes.

Examine the processed films for regions of higher optical density, which disclose the presence of poor contact between screen and film.

(b)   Equipment

Dedicated mammography screen–film contact test tool. Devices with a mesh of at least 20 lines/cm are suitable for this test. Those designed for general radiography are not suitable (Ardran *et al.*, 1969).

(c)   Results

Suspension: maximum of 4 cm^2 total area showing poor contact (except very close to identification window)

(d)   Additional information

The appearance of the dark patches is strongly influenced by the overall optical density of the film. Higher optical density tends to over-emphasise poor contact, which indicates the need to aim for a consistent optical density at each check. Contact is often poor near edges and particular attention should be paid to the chest wall edge. Poor contact is often observed close to the identification window, which is the only area where this can be tolerated. Regions of poor contact in the main area of the film may be due to the presence of dust or other particles between the screen and film, or to mechanical distortion. Particles can usually be identified by their characteristic appearance and cleaning the screen may remove the problem. Distortion is likely to require replacement of the cassette.

6.5.2   Sensitivity matching of cassette–screen–film batches

(a)   Method

A 4 cm thick Perspex block is placed on the table to cover the AEC detector completely, and the cassette placed in the Bucky. Select conventional AEC mode, 28 kV and a density setting to achieve the local target film optical density. Expose each cassette in turn, note the post-exposure mAs value and process immediately. Measure the densities on each film at a point on the film midline and 4 cm from the chest wall edge to correspond to the measurements of HVL and output (Sections 5.6.8, 5.6.9 and 5.7.3).

(b)   Equipment

4 cm Perspex phantom

Densitometer

(c)   Results

Remedial :   >0.1 variation in film density from the mean across the batch of cassettes

>5% variation in mAs from the mean across the batch of cassettes

(d)   Additional investigations

If it is required to investigate further, the procedure should be repeated using a fixed value of mAs to produce the local target film density. This will determine whether there are differences in the absorption properties of the cassettes.

Variations in the film density at constant mAs probably indicate variations in screen speed or sensitivity while variations in mAs indicate differences in the front or rear thickness of the cassette. Variations in mAs that are matched by variation in film density imply differences in the backs of the cassettes (including possibly the screens); those not so matched suggest differences in cassette fronts.

It is important that each x-ray unit has a set of matched cassettes, otherwise tracing of other problems may become difficult. In practice, the variations in density can quite easily be reduced to ±0.05 of the mean as suppliers can normally meet this level if asked to do so. Where a centre operates more than one mammography unit, it may be possible to sub-batch the cassettes so that individual mammography units are used with cassettes of very closely matched characteristics.

6.6   Darkroom and film storage conditions

The darkroom must be adequately light-tight to ensure that no extraneous light reaches the film before exposure and should contain no sources of light that would contribute to fogging of exposed films. Two methods of assessing darkroom light-tightness are presented here; the first is intended to be a quick subjective check undertaken more frequently than the second check, which is an objective measurement. Both checks should be undertaken routinely or as part of an investigation. It is anticipated that these checks would be undertaken by darkroom or radiographic staff on a routine basis. However, they may also be undertaken by physics staff as part of an investigation.

6.6.1   Subjective visual check

The person undertaking the check should ensure complete dark adaptation of the eyes by waiting in the darkroom for about 15 minutes. They should then inspect the darkroom for any areas of light ingress, particularly around the door, any louvres or ventilators and from around windows. Particular attention should be paid to indicator lamps. This interim check of the darkroom supplements the annual objective check and can be useful in detecting deterioration in light-proofing or the presence of any new unwanted sources of light.

6.6.2   Objective measurement

(a)   Method 1

Expose a mammography film inside a cassette to the local target optical density using 4 cm thick Perspex blocks. Place the film, emulsion side up, on the workbench or position where films are most likely to be handled. Place an opaque cover over one half of the film, widthways and leave on the film for two minutes. Remove the cover and process immediately. Repeat the process with the darkroom safelight switched on and under any other appropriate circumstances. Read the optical density of the exposed and unexposed halves of the film 4 cm in from the front edge and symmetrically about the midline of the film. Subtract one optical density value from the other.

(b)   Equipment

Opaque material sufficiently large to obscure half the film

Stopwatch with audible warning

4 cm Perspex blocks

Densitometer

(c)   Results

Remedial: >0.10 for a two minute exposure with the safelight illuminated (BSI, 1994b)

>0.025 for a two minute exposure with the safelight extinguished (BSI, 1994b)

Suspension: >0.20 for a two minute exposure with the safelight illuminated

>0.05 for a two minute exposure with the safelight extinguished

(d)   Method 2

Alternatively a method using two sensitometry stepwedges, one exposed on each short edge of the film, can be used. Here, one stepwedge is covered whilst the other is exposed. The difference in optical density between corresponding points on the two stepwedges, at around the target film density is recorded. If the wedges are exposed on the long edges of the film, an allowance may need to be made for optical density difference between the two stepwedges as a consequence of their orientation with respect to the direction of travel through the film processor.

(e)   Additional information

It is important that pre-exposed film is used for this test. In practice, film exposed to a midrange optical density value is significantly more sensitive to fogging than unexposed film. The shape of the characteristic curve means that a certain exposure to the film can be tolerated before there is any increase in the film base plus fog level. A similar exposure to a pre-exposed film will result in a much larger increase in density. The effect is illustrated in Figure 6.4 where the difference in optical density between two stepwedge films, one of which has been exposed in the darkroom for two minutes longer than the other, is shown.

Many darkrooms have tiny light leaks, noticeable only after dark adaptation for several minutes or when viewed from a particular direction. Provided they are not close to working areas, light leaks of this kind may be tolerated, but a 'soft glow' of daylight that appears after very little dark adaptation and is visible from any part of the darkroom is unacceptable.

Figure 6.4. Illustration of density shift observed between two stepwedge exposures, one exposed beneath a darkroom safelight for two minutes longer than the other

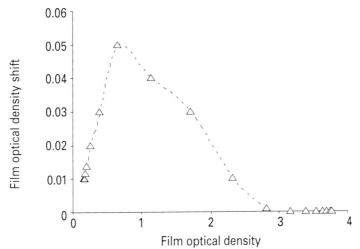

The safelights must also be tested to ensure that they are not responsible for increased fog level. Safelight filters such as Kodak GBX2 are standard for film sensitive to green light. Particular attention should be paid to the occurrence of cracks or defects in the filter material that may allow transmission of white light. The light bulb wattage should also be considered and should be no greater than 40 W (25 W if the distance is <1 m to the working surface).

Daylight processors and handling equipment are most frequently used. However, darkroom facilities are still essential for sensitometry and are highly desirable for use if problems arise with cassettes or film handling equipment. Some systems require that the film is loaded into magazines in a darkroom environment. At the planning stage it is therefore necessary to ensure that these facilities are properly provided. Louvres are often fitted in darkroom doors to provide ventilation and unless carefully designed will cause unacceptable light leaks. Some modern processors, designed primarily for daylight use, are installed into darkrooms particularly where space in the processing area is limited. These may have pilot lights or illuminated display panels generating sufficient light to render the darkroom unsafe for film handling. Mains socket pilot lights, if close to working areas or processor feeds and if orange rather than red, can also be a problem, as can indicator lamps on associated film processing equipment, computers, intruder alarm systems or smoke detectors.

6.7 Illuminators and viewing room conditions

Illuminators and viewing room conditions where mammograms are read should meet the standards described in MDA (1999b). Viewing boxes used only for checking mammograms for positioning need not meet these standards, but should be subject to the same checks in order to ensure stability of performance. Checks should be undertaken at least annually and the brightness of viewing panels measured. A further simple visual check should be carried out six monthly, which is normally undertaken by radiographers.

6.7.1 Subjective visual check

Examine the viewing panels visually. Any visible mismatches within or between light boxes in terms of brightness or colour should be attended to. Any visible flicker from the panels should result in the fluorescent tube being changed.

6.7.2 Objective measurements

(a)   Method

Switch on all the viewing panels and leave for at least five minutes to 'warm up' before any measurements are made. The luminance of the viewing panels can increase significantly during the first few minutes of operation. Measure the luminance at the centre of each panel and at positions towards each edge, not less than 20 mm from the edge of the panel, with the measuring probe positioned in accordance with the manufacturer's instructions. Measure the ambient light level (illuminance) in the room at the position of the film reader with the viewing boxes switched off.

(b)   Equipment

The device should be capable of making measurements of luminance up to 20,000 cd/m^2 (candelas per square metre) and measurements of illuminance up to 500 lux. Ideally, luminance measurements should be performed with a meter reading cd/m^2 and illumination by a detector reading in lux. If this is not possible then a cosine corrected illumination meter (capable of reading up to 70,000 lux) can be used to estimate the luminance. In this case the detector must be placed in direct contact with the viewing box panel and the illumination reading divided by π to give the luminance in cd/m^2. Whichever type of device is used, it is important that its spectral response should closely match that of the human eye.

(c)   Results

Remedial: luminance at the mid-point of each panel <3000 cd/m^2 (MDA, 1999b). This corresponds to an illumination value of approximately 9500 lux (MDA, 1999b) provided that the illumination has been measured with the detector in perfect contact with the viewing box panel

>15% variation between the luminance at the mid-points of each panel of an illuminator

>30% difference from the mid-point at other points within a panel and more than 20 mm from the edges of the panel (values outside this range indicate that the tubes may require replacement)

the illumination of the viewing room with viewing boxes switched off >50 lux (MDA, 1999)

(d)   Additional information

For viewing exceptionally dark areas in the radiographic image an additional spotlight should be available with a luminance of at least 20,000 cd/m^2 over an effective area of 8 cm diameter.

Studies have suggested that an important factor affecting the observer's ability to detect objects (in a test image) is reflection on the film (Robson, 2003). In a breast screening reporting room, the main sources of reflection are due to the illumination of the observer and other objects in the room by ambient light. The level of ambient light is governed in part by the amount of stray light emerging from around the edges of films mounted on the viewer. When an observer faces the viewer, this light illuminates them and their reflection is visible on the film. It may therefore be useful to check overall viewing conditions by simply observing a set of clinical films as they would normally be viewed by the film reader and noting any sources of light visible as reflection on the films. If the image of the observer is readily visible, this may indicate the need for improved masking of the films or reveal the presence of inadequately shuttered windows or inappropriately sited viewing boxes.

6.8   Image artefacts

Conventional film processing systems are prone to producing a variety of artefacts visible to a greater or lesser extent on the processed image. They may originate from the film processor itself, or any of the components of the daylight handling system with which the film may come into contact. Although most artefacts tend to arise within the processing system itself, the occurrence of image artefacts due to the automatic handling equipment should not be discounted when attempting to identify the origin of a problem. A discussion of film processing artefacts, their origin and diagnosis is presented in Haus and Jaskulski (1997). A method of producing test films to display artefacts is described here.

6.8.1   System artefacts

The films produced by the x-ray image uniformity test (see Section 5.7.2) may be used. Carefully examine the processed film, paying particular attention to the occurrence of film emulsion pick-off (white spots), roller marks (dark or light lines running parallel to the direction of film travel through the processor), scratch marks or drip marks, particularly at the leading and trailing edges of the film.

6.8.2   Film processing artefacts

A method of producing a film demonstrating artefacts arising solely from the film processing (Heid, 1998) may be preferred to the above method.

(a)   Method

In the darkroom, position a mammography film, emulsion side up, on a flat surface away from sources of light or reflective surfaces. Cover the film with a sheet of plain white paper. Make a spark using the cigarette lighter at a distance of approximately one metre, directly above the centre of the mammography film. Process the film and check that the film optical density falls within the range 1.0–2.0 above film base plus fog. Check the film optical density and examine for artefacts. Depending upon the type of cigarette lighter and the sensitivity of the film, more than one spark or adjustment of the distance may be required in order to achieve the required film optical density.

(b)   Equipment

Cigarette lighter with fuel removed

Sheet of plain white paper, large enough to cover mammography film

CHAPTER 7

Breast dose

7.1 Introduction

There is a risk of radiation-induced carcinogenesis associated with the x-ray examination of the female breast, but with modern equipment and technique this risk is small, whereas the benefit of the examination is considerable (NCRP, 1986). A number of authors have considered in particular the radiation risks and benefits of mammography in the context of breast screening (Beckett et al., 2003; Law, 1995, 1997; Law and Faulkner, 2001, 2002). Recently a working party of the National Radiation Protection Board and the NHS Breast Screening Programme reviewed the radiation risks of mammography and concluded that the lifetime risk of radiation induced breast cancer was about 1 per 100,000 per mGy in women screened in the UK (Young et al., 2003). Although this risk is relatively small it is important to use appropriate equipment and technique so that optimum image quality is achieved at the lowest possible dose consistent with that image quality. Thus the measurement of the dose to the breast forms an important part of the mammographic quality assurance programme.

The x-ray spectrum used for mammography is of low energy and the depth dose within the breast decreases rapidly with increasing tissue depth. For example, the exit dose measured by Hammerstein et al. (1979) for a 6 cm thick breast phantom varied between about 1% and 13% of the surface dose, depending upon beam quality. It is important therefore to specify breast dose using a quantity that gives a measure of the dose to the whole organ.

The dose specified in this document is the mean dose to the glandular tissues within the breast (Hammerstein et al., 1979; ICRP, 1987; NCRP, 1986). The term glandular tissue includes the acinar and ductal epithelium and associated stroma. It is the glandular tissues that are believed to be the most sensitive to radiation-induced carcinogenesis. The magnitude of the mean glandular dose will depend on the size and composition of the breast, with the former varying both within and between populations, and the latter throughout the life of the woman. Even when the average glandularity is the same, the distribution of glandular tissue is unpredictable and will vary from breast to breast. This variation makes dose estimation and comparison very difficult. For practical purposes, a standard breast model is used. In this revision a new standard breast model replaces the old one that assumed an average glandularity of 50%. This is because it has been realised recently that although a glandularity of 50% may be an appropriate assumption for breasts about 4 cm thick, it is not appropriate for thinner or thicker breasts, which will typically be more or less glandular respectively (Beckett and Kotre, 2000; Geise and Palchevsky, 1996; Klein et al., 1997; Young et al., 1998a). In this revision, additional factors are provided to allow for the modelling of variations in breast composition. In addition, many modern mammographic x-ray sets use target/filter combinations such as molybdenum/rhodium and rhodium/rhodium, which were not considered when the original calculations were made. This chapter presents data that extend the earlier tabulation of Dance (1990) to encompass glandularities of 0–100%, breast thicknesses of 2–11 cm and the x-ray spectra in current use. Other workers (e.g. Klein *et al.*, 1997; Wu *et al.*, 1991, 1994) have published similar factors. However, the approach presented here uses the same breast model as Dance (1990), and retains continuity with the dose prescription and *g*-factors given in the previous editions of this work (IPSM, 1989, 1994). Indeed, in many situations, the original factors can still be used without alteration.

7.2 Dose specification

The assessment of dose in mammography in the UK has traditionally been based on the estimation of the MGD for a 4.5 cm thick standard breast model with 50% glandularity in a central region. This was achieved by simulating the model with a 4.0 cm thickness of Perspex and applying a factor to estimate the entrance air kerma for the model. Recently some problems have arisen with this approach. One problem is that the average compressed breast for a typical screened population has a thickness of about 5.5 cm and a glandularity of about 30% (Dance *et al.*, 2000a; Young, 2002; Young and Burch, 2000). This means that average doses are rather higher than the MGD to the standard breast for a given system. Another issue is that on a modern mammography x-ray set the beam quality is selected automatically depending on breast thickness and composition, and the spectrum used may have significantly higher or lower energies. Consequently the use of a phantom comprising a 4.0 cm thickness of Perspex can result in an atypical spectrum being used to calculate the dose. To overcome these problems a new standard breast model is proposed. For clarity both models are described here.

7.2.1 Old standard breast model

The old standard breast model has a central region comprising a 50:50 mixture by weight of adipose and glandular tissue and a superficial region of adipose tissue, 0.5 cm thick. The compositions of these tissues are given by Hammerstein *et al.* (1979) and are well simulated by the tissue equivalent materials available from commercial suppliers (e.g. CIRS, USA). The old standard breast is 4.5 cm thick when compressed (Dance, 1990). It has a semi-circular cross-section in the horizontal plane of diameter 16 cm, which gives a cross-sectional area of approximately 100 cm^2. The exact area is not critical as the breast dose has only a small dependence on cross-sectional area (Dance, 1980).

The method for estimating the mean glandular dose uses a 4 cm thick Perspex phantom with the same cross-sectional area as the standard breast and conversion factors which have been specially calculated (Dance, 1990) to relate the incident air kerma for the Perspex phantom (which is readily measured) to the mean glandular dose for the standard breast. The mean glandular dose, D_{old}, for the old standard breast is then estimated using the relationship:

$$D_{old} = Kpgs \qquad (7.1)$$

where *p* converts the incident air kerma for the Perspex phantom (K) to that for the standard breast and *g* converts the incident air kerma for the standard breast to mean glandular dose. The conversion factors p and g are quality dependent and are given in Table 7.1. The factor s, which corrects for any difference from the original tabulation by Dance (1990) due to the use of a different x-ray spectrum, is given in Table 7.2 (Dance et al., 2000b).

Table 7.1. The conversion factors p and g for calculating the mean glandular dose to the old standard breast from measurements with a 4 cm thick Perspex phantom; these factors do not apply to the new standard breast

HVL (mm Al)	p	g (mGy/mGy)
0.25	1.12	0.155
0.30	1.10	0.183
0.35	1.10	0.208
0.40	1.09	0.232
0.45	1.09	0.258
0.50	1.09	0.285
0.55	1.07	0.311
0.60	1.06	0.339

Table 7.2. s-factors for clinically used spectra

Spectrum	s-factor
Mo/Mo	1.000
Mo/Rh	1.017
Rh/Rh	1.061
Rh/Al	1.044
W/Rh	1.042

7.2.2 New standard breast model

The data published in Dance et al. (2000a) for the equivalence between Perspex and typical compressed breasts has been used to establish a new standard breast simulated with a 4.5 cm thick Perspex phantom. The other dimensions are as for the old standard breast model. The entrance air kerma for a 4.5 cm thick Perspex phantom is equivalent to that for a 5.3 cm thick breast with a glandularity of 29% in the central region. The model still has 0.5 cm thick adipose layers at the top and bottom. The composition of this model was found to be typical for breasts of this compressed thickness for women in the age range 50–64. Note that a p-factor is no longer necessary. The mean glandular dose, D_{new}, for the new standard breast is estimated using the relationship:

$$D_{new} = K_{45} \cdot g_{53} \cdot c_{53} \cdot s \quad (7.2)$$

where, K_{45} is the entrance air kerma for 4.5 cm thickness of Perspex, $g_{53}=g$-factor for the 5.3 cm thick standard breast, c_{53} is the conversion factor which allows for the glandularity of the 5.3 cm thick standard breast, and the s factor is the spectral correction factor. The g and c factors are dependent on the HVL of the spectra used and estimated from Table 7.3. These factors are explained further in Section 7.4.

Table 7.3. The conversion factors g and c for calculating the mean glandular dose to the new standard breast (5.3 cm thick) from measurements with a 4.5 cm thick Perspex phantom

HVL (mm Al)	g (mGy/mGy)	c	Product of g and c
0.30	0.155	1.109	0.172
0.35	0.177	1.105	0.196
0.40	0.198	1.102	0.218
0.45	0.220	1.099	0.242
0.50	0.245	1.096	0.269
0.55	0.272	1.091	0.297
0.60	0.295	1.088	0.321

7.3 Estimation of mean glandular dose for the standard breast

The mean glandular dose to the new standard breast is estimated using the Perspex phantom described in Section 7.2.2. The Perspex thickness should be within 0.05 cm of its nominal thickness of 4.5 cm (Faulkner et al., 1995). A loaded cassette should be placed in the cassette holder. The phantom should be exposed using the settings normally used in clinical practice for a breast of size and composition similar to the 5.3 cm thick standard breast. This can usually be achieved by allowing the x-ray set to make an automatic selection of beam quality and mAs. (The phantom may also be exposed and dose calculated at standard settings (28 kV Mo/Mo) for purposes of intercomparison and investigation.)

The method involves a determination of both the incident air kerma for correct exposure of this phantom and the half value layer of the x-ray beam. Measurement of the former quantity is divided into two stages: a determination of the mAs required for the correct exposure of the Perspex phantom and a measurement of tube output per mAs. Methods for the measurement of tube output and of HVL are described in Sections 5.6.9 and 5.6.8 respectively. Both these quantities should be measured with the compression plate in place.

The mAs per exposure is determined using the Perspex phantom, which is positioned as for a cranio-caudal exposure of the breast with the compression device in place against its top surface. It is advisable to use a standard compression force (e.g. 100 N) as on some systems the exposure may be affected by the applied force. Some x-ray sets automatically choose the beam quality depending on the compressed breast thickness. In this case it may be more accurate to use an expanded polystyrene spacer to increase the thickness to 53 mm during compression. Care should be taken to avoid covering the AEC detector with the spacer. The position of the AEC detector should also be standardised using the location closest to the chest wall edge and on the midline. Expose and process a film and check that the resulting optical density is within the normal range. Alter the exposure time or density control if necessary. Record the mAs per exposure (mAs_{exp}).

For this mAs value, measure the air kerma directly in the plane equivalent to the top surface of the breast phantom (but in the absence of the phantom). Alternatively measure the distance from the focal spot of the x-ray tube to the top surface of the breast phantom (d cm) and deduce the air kerma incident on this phantom per exposure using the tube output (air kerma per mAs at 50 cm, T, from Section 5.6.9), the mAs per exposure (from above) and the inverse square law. Include attenuation by the compression plate in the estimation. Do not include the effect of backscatter on the air kerma measurement as this is accounted for in the original calculation of g-factors. Note that the output should be measured on the axis specified in Section 5.6.9(b), i.e. on the midline and 4 cm from the chest wall edge of the film. The measurement of air kerma should be made with the paddle positioned so that there is a gap of at least 5 cm between the paddle and the dosimeter.

$$\text{Air Kerma (K)} = T \cdot \text{mAs}_{exp} \cdot (50/d)^2 \tag{7.3}$$

The mean glandular dose to the standard breast is calculated using Equations 7.2 and 7.3. The correction factors g and c at the measured half value layer may be interpolated from the values given in Table 7.3. When quoting measured dose it is essential to specify both the dose quantity and the breast size and composition that have been assumed. The new standard breast model can be expected to result in doses that are approximately 25% greater than those found with the old model if the same beam quality (28 kV Mo/Mo) is used.

Equipment:	Perspex phantom, 4.5 cm thick
	ionisation chamber/electrometer
	densitometer
	aluminium foils (i.e. at least 99.9% purity)
	suitable jig
Remedial value:	the mean glandular dose for the new standard breast at clinical settings >2.5 mGy/film

Values less than 1.0 mGy/film should be investigated to decide whether the image quality obtained is adequate.

The method described above is appropriate at the commissioning of the x-ray set and for infrequent checks of breast dose, i.e. six monthly. For frequent checks of breast dose, the results of the frequent AEC checks described in Section 5.7.3(b) are sufficient. It is recognised that this procedure only gives an indirect indication of breast dose, but a more detailed method is not practicable for a frequent test.

7.4 Estimation of mean glandular dose to real breasts

It is recommended to periodically measure the mean glandular doses for a series of breast examinations on each mammographic system (IPEM, 1997). To do this, the breast thickness under compression is measured, and the tube voltage, and mAs delivered are recorded.

From a knowledge of the output of the x-ray set for the kV and target and filter material used, this mAs value may be used to estimate mean glandular dose using the expression:

$$D = Kgcs \qquad (7.4)$$

where K is the incident air kerma calculated (in the absence of scatter) at the upper surface of the breast. The factor g corresponds to a glandularity of 50%, and is the same as that given in IPSM (1994) with an extension of the table to cover breasts in the thickness range 9–11 cm as shown in Table 7.4. The factor c corrects for any difference in breast composition from 50% glandularity. C-factors for typical breast compositions in the age ranges 50–64 and 40–49 are shown in Tables 7.5 and 7.6. There is an error associated with using these tables for other age groups or individual breasts where glandularity may be different from the average. C-factors tabulated in Dance et al. (2000a) as a function of glandularity may be useful in this situation. The factor s corrects for any difference due to the choice of x-ray spectrum as noted earlier. Measurement of compressed breast thickness for this purpose is performed by the radiographer, by reading the displayed compressed thickness on the x-ray set. The accuracy of the displayed thickness should be verified by applying a typical force (e.g. 100 N) to rigid material of known thickness. It may be necessary to apply correction factors if the displayed values are in error. An accuracy of better than ±0.2 cm is required. Software for making such dose calculations has been published by the NHSBSP (Young, 2001).

Table 7.4. g factors (mGy/mGy) for breast thicknesses of 2–11 cm and the HVL range 0.30–0.60 mm Al; the g-factors for breast thicknesses of 2–8 cm are taken from Dance (1990), and for 9–11 cm from Dance et al. (2000a)

Breast thickness (cm)	HVL (mm Al)						
	0.30	0.35	0.40	0.45	0.50	0.55	0.60
2	0.390	0.433	0.473	0.509	0.543	0.573	0.587
3	0.274	0.309	0.342	0.374	0.406	0.437	0.466
4	0.207	0.235	0.261	0.289	0.318	0.346	0.374
4.5	0.183	0.208	0.232	0.258	0.285	0.311	0.339
5	0.164	0.187	0.209	0.232	0.258	0.287	0.310
6	0.135	0.154	0.172	0.192	0.214	0.236	0.261
7	0.114	0.130	0.145	0.163	0.177	0.202	0.224
8	0.098	0.112	0.126	0.140	0.154	0.175	0.195
9	0.0859	0.0981	0.1106	0.1233	0.1357	0.1543	0.1723
10	0.0763	0.0873	0.0986	0.1096	0.1207	0.1375	0.1540
11	0.0687	0.0786	0.0887	0.0988	0.1088	0.1240	0.1385

Table 7.5. c-factors for average breasts for women in age group 50–64 (Dance et al., 2000a)

Breast thickness (cm)	HVL (mm Al)						
	0.30	0.35	0.40	0.45	0.50	0.55	0.60
2	0.885	0.891	0.900	0.905	0.910	0.914	0.919
3	0.925	0.929	0.931	0.933	0.937	0.940	0.941
4	1.000	1.000	1.000	1.000	1.000	1.000	1.000
5	1.086	1.082	1.081	1.078	1.075	1.071	1.069
6	1.164	1.160	1.151	1.150	1.144	1.139	1.134
7	1.232	1.225	1.214	1.208	1.204	1.196	1.188
8	1.275	1.265	1.257	1.254	1.247	1.237	1.227
9	1.299	1.292	1.282	1.275	1.270	1.260	1.249
10	1.307	1.298	1.290	1.286	1.283	1.272	1.261
11	1.306	1.301	1.294	1.291	1.283	1.274	1.266

Table 7.6. c-factors for average breasts for women in age group 40–49 (Dance et al., 2000a)

Breast thickness (cm)	HVL (mm Al)						
	0.30	0.35	0.40	0.45	0.50	0.55	0.60
2	0.885	0.891	0.900	0.905	0.910	0.914	0.919
3	0.894	0.898	0.903	0.906	0.911	0.915	0.918
4	0.940	0.943	0.945	0.947	0.948	0.952	0.955
5	1.005	1.005	1.005	1.004	1.004	1.004	1.004
6	1.080	1.078	1.074	1.074	1.071	1.068	1.066
7	1.152	1.147	1.141	1.138	1.135	1.130	1.127
8	1.220	1.213	1.206	1.205	1.199	1.190	1.183
9	1.270	1.264	1.254	1.248	1.244	1.235	1.225
10	1.295	1.287	1.279	1.275	1.272	1.262	1.251
11	1.294	1.290	1.283	1.281	1.273	1.264	1.256

7.5 Simulation of different thicknesses of breasts using Perspex

It is possible to estimate the doses to a range of thicknesses of breasts of typical composition using blocks of Perspex as a breast phantom. This method relies on the equivalence between different thicknesses of Perspex and typical breasts published by Dance et al. (2000a) and summarised here in Table 7.7. It should be noted that since Perspex is generally denser than breast tissue any automatic selection of kV, target or filter may be slightly different from real breasts. This can be corrected by adding expanded polystyrene blocks to the Perspex as a spacer to make up a total thickness equal to the equivalent breast. Care should be taken to avoid the area of the AEC detector. On systems that determine the exposure factors primarily on attenuation, such as the GE Senographe DMR, such spacers should not be necessary. The mean glandular dose (D) to a typical breast of thickness and composition equivalent to the thickness of Perspex tested is calculated by applying Equation 7.4. In this case, K is the incident air kerma at the upper surface of the Perspex block. The c- and g-factors applied are those for the corresponding thickness of the typical breast rather than the thickness of Perspex block as shown in Tables 7.7 and 7.8.

Table 7.7. c-factors for breasts (age 50–64) simulated with Perspex, assuming surface layers of adipose tissue 0.5 cm thick (Dance et al., 2000a)

Perspex thickness (cm)	Equivalent breast thickness (cm)	Glandularity of equivalent breast (%)	HVL (mm Al)						
			0.30	0.35	0.40	0.45	0.50	0.55	0.60
2.0	2.1	97	0.889	0.895	0.903	0.908	0.912	0.917	0.921
3.0	3.2	67	0.940	0.943	0.945	0.946	0.949	0.952	0.953
4.0	4.5	41	1.043	1.041	1.040	1.039	1.037	1.035	1.034
4.5	5.3	29	1.109	1.105	1.102	1.099	1.096	1.091	1.088
5.0	6.0	20	1.164	1.160	1.151	1.150	1.144	1.139	1.134
6.0	7.5	9	1.254	1.245	1.235	1.231	1.225	1.217	1.207
7.0	9.0	4	1.299	1.292	1.282	1.275	1.270	1.260	1.249
8.0	10.3	3	1.307	1.299	1.292	1.287	1.283	1.273	1.262

Table 7.8 g-factors (mGy/mGy) for breasts simulated with Perspex

Perspex thickness (cm)	Equivalent breast thickness (cm)	HVL (mm Al)						
		0.30	0.35	0.40	0.45	0.50	0.55	0.60
2.0	2.1	0.378	0.421	0.460	0.496	0.529	0.559	0.585
3.0	3.2	0.261	0.294	0.326	0.357	0.388	0.419	0.448
4.0	4.5	0.183	0.208	0.232	0.258	0.285	0.311	0.339
4.5	5.3	0.155	0.177	0.198	0.220	0.245	0.272	0.295
5.0	6.0	0.135	0.154	0.172	0.192	0.214	0.236	0.261
6.0	7.5	0.106	0.121	0.136	0.152	0.166	0.189	0.210
7.0	9.0	0.086	0.098	0.111	0.123	0.136	0.154	0.172
8.0	10.3	0.074	0.085	0.096	0.106	0.117	0.133	0.149

7.6 Additional factors affecting mean glandular dose

7.6.1 The consequences of multiple films or subsequent examinations

Preceding sections have been concerned with the mean glandular dose for a single film covering the whole of the breast. Estimates of risk should be based on the average of the doses to the two breasts. For a given projection it is usual to take one film of each breast examined. In practice, if a breast is too large to be imaged on a single 18×24 cm film, two (or occasionally more) films will be required per view. The dose for each exposure can be estimated using that for irradiation of the whole breast multiplied by the fraction of the breast irradiated. The procedure for dose estimation has usually been simplified by assuming that the entire breast is included with each exposure and the doses per film added together (Burch and Goodman, 1998; Young and Burch, 2000). This results in an estimate of the mean glandular dose, which is proportional to the number of films taken. If more than one view of the same breast is taken, such as medio-lateral oblique and cranio-caudal, then the mean glandular dose for each view should be calculated separately and added together. The mean glandular dose for medio-lateral oblique views is on average greater than for cranio-caudal views (Burch and Goodman, 1998; Young and Burch, 2000).

7.6.2 Effect of magnification

When magnification views are taken, the image receptor is at the same focus-film distance as in non-magnification mammography or at a slightly greater distance. It is common practice to radiograph the breast on a platform some distance away from the image receptor. The introduction of an air-gap, which reduces the effect of scattered radiation on the image receptor, obviates the need for an anti-scatter grid. The incident air kerma to the breast increases with the reduction in focus skin distance due to the inverse square law. However, the g-factor (Equation 7.2) is little affected by changes in focus skin distance. If the whole of the breast is included in the irradiated field and the grid is removed, the dose for magnification is approximately three times that for a contact film. In practice most magnification films are taken with the primary beam restricted to the region of clinical interest so that the mean glandular dose is comparable with (and may be less than) that for a contact film. Where doses are to be estimated for magnification films of part of the breast any of the above methods may be used with the introduction of an area factor (a). Multiplication by the a-factor corrects for the reduced area of the compressed breast that is exposed, calculated as in Equation 7.5 for a standard breast 5.3 cm thick:

$$a = \frac{\text{area irradiated at breast platform} \left(\frac{FTP \pm 2.65}{FTP}\right)^2}{\text{area of compressed breast}} \quad (7.5)$$

where FTP is the focus-to-platform distance in cm. If the area of the compressed breast is not known, the calculation may be simplified by using the area of the standard breast, 100 cm^2. In any event the a-factor should not exceed 1.

Due to beam divergence the irradiated area of the breast depends on the plane of the breast chosen. Taking the area at the mid-plane of the breast provides a reasonable compromise between extremes. The area irradiated at the breast table may be determined by exposing a cassette on the magnification table. The area of the standard breast is 100 cm^2. (Note that this is a standard assumption and does not reflect average breast area, but does allow for a simple and approximate correction for the reduced area exposed in reduced field imaging.)

7.6.3 Stereotactic localisation

The dose for images taken for stereotactic localisation may be estimated in a similar manner to that described for magnification using the area factor described in Equation 7.5. In this case FTP represents the distance from the focus to the breast support table rather than the magnification platform. Note that in this case the value used for the output and HVL should not include the effect of transmission through a compression paddle. The effect of the angle of the incident x-ray beam may be ignored for simple calculations. Stereotactic procedures usually involve two exposures on one breast. Comparison of the radiation risk of stereotactic procedures with standard mammographic imaging should be based on the average of the dose to the two breasts. There is currently no guidance on acceptable doses for stereo systems. The procedure described here can be applied to either screen–film or digital stereotactic systems.

CHAPTER 8

Image quality and test objects

8.1 Introduction

In mammography the quality of the image is of prime importance. The consequences of poor image quality are twofold: the radiologist may not have all the diagnostic information that should be available and the patient may have received unnecessary radiation exposure, particularly if a repeat film is necessary. A balance must be maintained between the best achievable image quality and radiation dose to the patient. Any exposure is wasted if it is insufficient to produce a diagnostically useful mammogram. Adequate image quality must be obtained with a dose as low as reasonably practicable. In early guidance on the establishment of a quality assurance system for mammography used for breast screening (DH, 1989a) the first objective and standard was 'to achieve optimum image quality'. This priority has not changed: current assessments of benefit versus risk in the breast screening programme indicate a very significant advantage in benefit over risk at current dose levels and age of screened population (Law and Faulkner, 2001). Thus the objective of achieving optimum imaging performance necessitates an assessment of image quality as well as dose during the commissioning and routine testing of a mammographic x-ray system.

8.2 Clinical image quality

It is important to define what type of image is required and what information is required from it. This may vary from patient to patient and from radiologist to radiologist. It may also depend on whether the mammogram is for symptomatic diagnosis, screening, assessment or whether specific localisation or other mammographic information is requested. In screening for example, image quality must provide high sensitivity for invasive carcinomas under 1 cm in diameter (DH, 1989a).

The interpretation of mammograms is complicated by the changes in breast composition and structure with age (Parsons, 1983). The normal breast consists of fibrous, glandular and adipose tissue. As the woman ages, and following pregnancy or the menopause, the fibrous and glandular tissue is replaced by fat. Cysts, fibrous tumours and dilated ducts are all benign conditions that may occur during these changes. The image quality required for the visualisation of a large mass in a fatty breast may be very different from that required to show small calcifications in a more glandular breast.

In general, lesions of the breast are indicated by masses, calcifications and/or distortions of breast architecture. A positive diagnosis will depend not only on their presence in the radiographic image but also on their number, size, shape and configuration. The following clinical features are important:

Calcifications. These are composed of calcium hydroxyapatite and/or tricalcium phosphate. Calcifications with a size greater than 0.1 mm may be detected in mammograms. They are most often irregularly shaped and present in clusters. Approximately half of screen detected cancers show calcifications in screening mammograms, and in half of these a calcification cluster was the sole diagnostic feature of the cancer (Meeson and Young, 2001). An accurate representation of the shape of the calcifications is regarded as an important aid to diagnosis (Evans A *et al.*, 2002). A single calcification is unlikely to be diagnostic.

Masses. Soft tissue masses may vary in size from a few millimetres to several centimetres. If cancerous they are usually irregularly shaped and often infiltrate into the surrounding tissues with fine spicules. Masses have approximately the same density as glandular tissue and are denser than adipose tissue. For example, a 3 mm infiltrating duct carcinoma in a 42 mm thick breast is calculated to have a subject contrast of only 1.3% (Johns and Yaffe, 1987), which is close to the limit of detectability with present screen–film systems.

Skin thickening, engorged ducts and distortion of breast architecture. These are other signs that may be recognised in the mammogram as indicative of disease.

Image quality is therefore concerned primarily with the ability of the system to demonstrate the very small changes in soft tissue contrast and to detect the calcifications that are also often associated with breast disease. However, by virtue of their size and contrast, these critical features are difficult to detect in the presence of the structured noise associated with normal breast architecture. A good image is therefore characterised by the ease with which critical features can be seen in the presence of this noise.

There has been an attempt at specifying clinical image quality criteria. A European Study Group has recommended that the mammographic image should have visually sharp reproduction of the whole glandular breast, cutis and subcutis (CEC, 1996). For good imaging performance the visibility of 3 mm diameter round details and 0.2 mm microcalcifications are listed as being important.

8.3 Image quality assessment

Image quality depends upon many factors and interactions between components of the imaging process. Physical aspects that have a particular bearing on image quality are discussed in Chapter 3.

Modulation transfer function (MTF) can give a useful measure of system sharpness properties (ICRU, 1986, 1996). MTF has been applied in mammography (Bencomo *et al.*, 1982) but it is not a measure of overall performance (Moy, 2000). Other quantitative measures such as the noise power spectra (NPS) need to be considered along with the MTF to fully characterise the overall performance of the unit in terms of the signal-to-noise ratio. Although these techniques have been used to assess screen–film systems (Bunch, 1997; MDA, 1998), they are more often used when assessing digital imaging systems (Evans DS *et al.*, 2002; Maidment and Yaffe, 1994; Thunberg *et al.*, 1999; Vedantham *et al.*, 2000; Workman *et al.*, 1994). Although useful, such measurements are generally considered as laboratory work and are unlikely to be applied in a routine quality assurance programme. Receiver operating characteristic (ROC) analysis, which includes observer performance in the assessment of overall performance, has also been used in mammography (Moores et al., 1979; Säbel et al., 1986; Yip et al., 2001). However, due to time requirements ROC analysis is impracticable for the routine assessment of image quality.

Although many individual physical parameters affecting image quality can be measured in a test situation, the pragmatic approach traditionally used in mammography is to attempt to measure overall imaging performance using an image quality test object.

8.4 Image quality test objects

Image quality test objects are designed to be imaged under clinical conditions to simulate the mammographic examination. They contain test details and in some cases may resemble the shape and composition of the compressed breast. They can be used to provide a quality control check on the variation of performance of a particular mammographic system, or to assess the performance of one mammographic system against another or against recommended guidelines.

Test objects have been in use since the early days of mammography. The ideal image quality test object should be easy to use, and be suitable for use with automatic exposure control to simulate the clinical exposure. Test details should be clinically relevant and be sufficiently sensitive to register the small changes seen in system performance, particularly those relating to spatial resolution and contrast detectability. Quantitative evaluation of image quality parameters should be possible and the test object should ideally facilitate objective rather than subjective analysis so that image quality may be compared irrespective of observer performance or experience. It is assumed that better images of test objects correspond with better diagnostic images, but this depends on the nature of the details and background structures.

Image quality test objects can be divided into the following categories:

(i) 'Simple' test objects providing a non-quantitative subjective assessment of image quality. These contain a variety of test detail which may resemble breast tissue or represent other breast features, e.g. microcalcifications.

(ii) 'Scoring' type test objects in which the test details visible are assigned a numerical value or counted to provide a quantitative assessment of image quality. These may contain an assortment and range of test details such as discs, fibres and particles that simulate clinical features including masses, spicules and calcifications.

(iii) 'Objective' test objects that contain test details relevant to the measurement of particular physical indices such as spatial resolution or contrast detectability. Such details are likely to include defined contrasts for a range of sizes together with calibrated resolution test patterns. There remains subjectivity in the use of these test objects.

Many of the test objects that have been produced commercially do not relate well to the ideal, nor do they fit readily into one or other of the above categories. Some test objects actually combine two or all three of the above functions, in an attempt to increase their versatility. Simple test objects in category (i) can provide a rapid check on system performance, particularly when comparing images side by side, and this may satisfy the need for frequent image quality tests by radiographic staff (NHSBSP, 1999b). Anthropomorphic test objects would fall into this category.

Those in category (ii) that involve evaluating visible details to obtain an image quality 'score' constitute the majority of presently available test objects. The use of such test objects weights different aspects of image quality in a rather arbitrary way. The scoring methodology can be laid down but there is still variation of observer performance and difficulties in ensuring that scoring criteria are consistently applied (Faulkner and Law, 1993; Young and Ramsdale, 1993). Interpretation and cross-comparison of results can therefore be difficult (Law, 1991; Moores, 1991; Robinson and Underwood, 1991). Observer performance may be less of a problem if simple detection criteria are defined and universally adopted.

Test objects in category (iii) that attempt to measure accepted image quality indices such as limiting high contrast spatial resolution and minimum perceptible contrast are more likely to satisfy the requirement for a more sophisticated test object. With a 'calibrated' test object, where the characteristics of the test detail are known, there is the possibility of making objective measurements, which can be directly compared, from system to system. Some test objects in this category (Ramsdale, 1989; Thijssen et al., 2001; Thompson and Faulkner, 1991) utilise contrast detail for a range of sizes enabling the generation of a contrast detail or threshold contrast detail-detectability diagrams. The curve of threshold contrast against size is taken to be one measure of overall performance. Threshold contrast is assessed by measuring the lowest detectable contrast of circular detail as a function of the detail area. Threshold contrast is affected by various factors including the sensitivity of the imaging system, structural and random noise and image unsharpness.

A suitable test object for measuring threshold contrast must have detail sizes and a contrast range appropriate for mammography. Test objects with suitable contrasts and detail sizes include the TOR(MAX) (Cowen and Coleman, 1990) and the CDMAM (Thijssen et al., 2001). It is important that the contrast difference between neighbouring details is fine enough to test the imaging system adequately.

Threshold contrast data may be presented graphically as the threshold detection index (H_T) for a range of detail sizes defined as in Equation 8.1. Measured data on detection indices for typical analogue and digital mammography systems using the CDMAM are shown in Figure 8.1. The same data are presented as a contrast detail diagram in Figure 8.2.

$$H_T(A) = 1/(C_T \cdot A^{0.5}) \qquad (8.1)$$

where C_T = threshold contrast, A = detail area.

It is important to maintain a strict scoring protocol, i.e. fixed or variable distance scoring, and to maintain constant threshold criteria.

Figure 8.1. Detection index [H(t)] for typical digital and analogue mammography systems

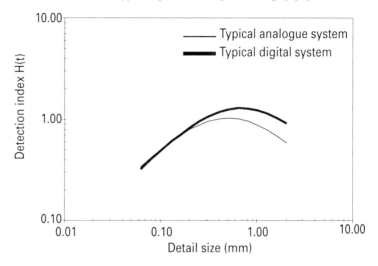

Figure 8.2 Contrast detail diagram for typical analogue and digital mammography systems

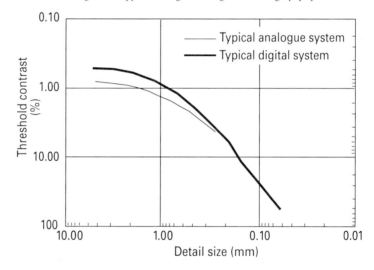

Alternatively, the threshold contrast for certain detail sizes, which relate to clinically significant details can be assessed. For screen–film mammography systems, the National Health Service Breast Screening Programme (NHSBSP) uses detail sizes of 5–6 mm, 0.5 mm and 0.25 mm, which should have threshold contrasts of better than 1.2%, 5% and 8% respectively (NHSBSP, 2004). If a single index is required, there have been several suggestions for a suitable measure, e.g. an 'image quality figure' derived from measurements of contrast versus detail size (Thijssen et al., 1989).

A major difficulty with many of the early test objects of type (ii) or (iii) was a lack of suitable range and sensitivity when compared with the performance of a modern mammographic system. The more recently designed image quality test objects reveal an increased awareness of the required sensitivity. In general, more sensitive test objects require more care and concentration in scoring and results may vary more widely between users. However, a test object, which shows little change between good and poor imaging systems, is of little use.

The TOR(MAX) test object (see Figure 8.7) has been used by most centres in the breast screening programme in the United Kingdom. A second test object designed by the same group, the TOR(MAM) (Cowen et al., 1992) (see Figure 8.8) is also used by most centres. Mammography programmes in other countries utilise other test objects such as the RMI, CIRS and CGR(IGE). Reports of measurements of image quality in the NHSBSP using the TOR(MAX) and TOR(MAM) are included in reviews of mammographic equipment and its performance in the NHSBSP (Young et al., 1992, 1998b, 1998c).

8.5 Test object use

The following general considerations apply to using test objects to investigate image quality.

8.5.1 Standard factors

It is important to decide whether one is testing a mammography system as used clinically or how it performs under standard conditions. Both approaches may be valid under different circumstances. Some of the major factors are discussed below.

(a) Beam quality

It is now modern practice to use different beam qualities for different types of breast. It is not practical other than in a research situation to explore all of the available beam qualities. One could allow the system to automatically select a beam quality using the same settings as with a patient. However this will modify the beam quality in a way that is not necessarily representative of patients and may make comparison between systems difficult. It is suggested therefore that a measurement of image quality should always be conducted using standard factors of 28 kV with a Mo/Mo target/filter combination. In certain situations further measurements may be desirable using other beam qualities. Certain test objects have details with a calibrated contrast, which are only accurate for these standard exposure factors (e.g. TOR(MAX)).

(b) Film density

Image quality is strongly affected by the background density of the test object film. It is usual practice to expose the test object using AEC. This should then produce densities comparable to those found in clinical practice. However, care should be taken that the exposure is not unduly affected by any details that lie within the sensitive area of the AEC detector. In a well-designed test object high contrast details are positioned away from this central area. It is good practice to measure the background film density of the film of the test object at some standard point. When using the TOR(MAX) and

TOR(MAM) test objects this should be at 4 cm from the chest wall edge in a uniform area. Ideally this density should be consistent with local clinical practice and consistent over time. If the current clinical practice or a fault in the x-ray set results in an unusually low or high density, a further film should be produced with a background density closer to the target density (see Section 5.7.3 (b)(i)).

(c)   Scatter material

Some test objects comprise a relatively thin test plate that is designed to be used with additional scatter material. The amount of scatter material can strongly affect the appearance of test films due to the effect of scatter, beam hardening and geometric blurring. It is therefore important to standardise on the amount used. An additional 3–4 cm of plastic material (usually Perspex) is usually placed below the test plate. It is standard practice to use an additional 3 cm of Perspex with the TOR(MAM) and 4 cm of Perspex with the TOR(MAX).

(d)   Screen–film cassette

Image quality measurements are generally made using a particular cassette and film taken from the local stock. It is common practice to use the radiographers' test cassette (QA cassette). It is recommended that the same cassette be used for all measurements on a particular system.

8.5.2   System resolution versus detector resolution

Image quality test objects are usually used to measure the total mammography system. However in some cases one may wish to measure the performance of the detector system in isolation. Thus, for example, one may measure the limiting spatial resolution of a screen–film system by imaging a test grating placed directly onto the cassette. In this case there is no significant geometric blurring and the resolution measured is that of the screen–film detector. When measuring system resolution the grating is placed on top of a stack of Perspex, typically 4 cm thick. The presence of geometric blurring, and to a lesser extent scatter, results in a measurement of resolution that is lower than that of the screen–film detector alone. Note that the limiting resolution of modern screen–film systems may exceed the maximum (20 line pairs per mm) that can be measured with commonly available test gratings.

8.5.3   Viewing conditions

The quality of viewing conditions can affect the reading of test object films. The general requirements are similar to those described for clinical films (see Section 6.7). The film should be masked to minimise glare and an adequately bright viewing box should be used. The ambient light level should be low (e.g. less than 50 lux) and reflections minimised. Magnifying glasses should be used where appropriate.

8.6 Image quality standards

Where image quality standards have been defined they have usually been applied to the high contrast limiting spatial resolution and threshold contrast detection. Standards for image quality in the UK have previously been defined in two documents. Standards for the equipment used by the NHSBSP are summarised in NHSBSP (2005). Image quality standards for mammography are also included in guidelines covering the whole of diagnostic radiology in the UK (IPEM, 1997). Since this latter document was written there have been significant improvements in the performance of mammographic equipment. Most noticeable has been the improvement in high contrast spatial resolution with the development of smaller focal spot sizes. There has also been some improvement in the minimum threshold contrasts detected with better screen–film combinations. The measurement of image quality is subjective and due allowance should be made for observer variability. Although no attempt has been made to limit the standards to a particular test object design, they are based on measurements using the TOR(MAX) test object. The standards are not regarded as appropriate for digital mammography and new standards and testing procedures for such systems are in preparation.

8.6.1 Limiting high contrast spatial resolution

(a) Introduction

Limiting high contrast spatial resolution is an important performance parameter for screen–film systems. It is sensitive to the geometric blurring of enlarged focal spots, and differences in screen resolution. The results of high contrast spatial resolution measurements using the TOR(MAX) test object for 270 mammography systems in the NHSBSP are shown in Figure 8.3 (Young *et al.*, 2000).

(b) Measurement

The measurement of limiting high contrast spatial resolution is affected by the positioning of the resolution grating. The resolution measurement should be on top of 4 cm of Perspex and approximately 6 cm from the chest wall edge and in directions both parallel to and perpendicular to the tube axis. The measured resolution in the direction perpendicular to the tube axis will normally be greater than that parallel to the tube axis because of the asymmetrical nature of the focal spot. The standard should be met in both directions. The test film should have sufficient dwell time in the cassette prior to exposure to ensure good screen–film contact. The film should be evaluated under appropriate conditions with the aid of a high power magnifier.

Equipment: high contrast resolution test grating (up to at least 20 line pairs per mm); this can be used separately or as part of a test object (e.g. TOR(MAX))

4 cm thick Perspex

×10 magnifying glass

Figure 8.3. High contrast spatial resolution measurements using TOR(MAX) or TOR(MAS) test objects in the NHSBSP (Young et al., 2000)

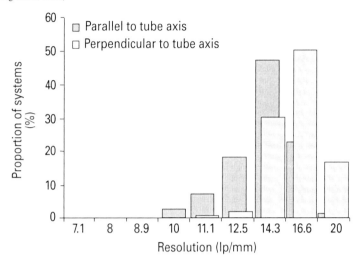

Remedial: <12 line pairs per mm visible

>25% change in line pairs per mm from baseline value (for analogue systems only)

Suspension: <10 line pairs per mm visible

8.6.2 Threshold contrast

(a) Introduction

Threshold contrast is most commonly measured with a TOR(MAX) test object. Note that the nominal contrast of each detail published by the manufacturer has been calculated for an exposure using 28 kV and a Mo/Mo target/filter combination. The nominal contrasts quoted by the manufacturer and in the remedial and suspension levels used here do not include the reduction in contrast due to scatter. The use of an alternative beam quality changes the contrast of the details and would invalidate the measurements at least in terms of threshold contrast. The results of threshold contrast measurements using the 0.25 mm, 0.5 mm and 6 mm details in the TOR(MAX) test object for mammography systems in the NHSBSP are shown in Figures 8.4–8.6.

(b) Measurement

The test object should be placed on top of the 4 cm of Perspex and imaged using a tube voltage of 28 kV and a Mo/Mo target/filter combination under AEC control. The density of the film should be tested to ensure that it is consistent with local clinical practice, as discussed in Section 8.5.1(b). The lowest contrast that can be detected for each detail size should be determined by viewing the film on a masked illuminator with a low level

Figure 8.4. Threshold contrast measurements for 0.25 mm details for mammography systems in the NHSBSP (Young et al., 2000)

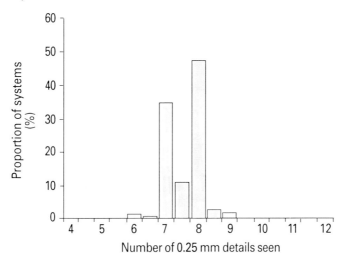

Figure 8.5. Threshold contrast measurements for 0.5 mm details for mammography systems in the NHSBSP (Young et al., 2000)

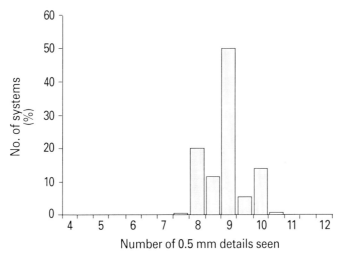

of ambient light. Test films may be read unaided or using a magnifier of the type favoured by radiologists. In cases of doubt additional films should be taken and more than one film reader involved. The average minimum detectable contrast should be determined.

Equipment: a test object with details of appropriate diameters and a range of known contrasts at 28 kV Mo/Mo (e.g. TOR(MAX))

a 4 cm thickness of Perspex

×2 magnifying glass (i.e. as used by radiologists)

Figure 8.6. Threshold contrast measurements for 6 mm details for mammography systems in the NHSBSP (Young *et al.*, 2000)

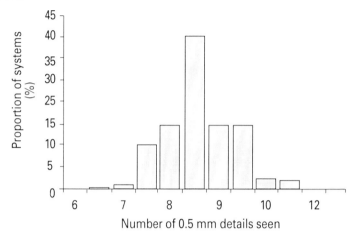

Detail diameter	Target	Minimum detectable contrast Remedial	Suspension
5–6 mm	>0.8%	>1.2%	>1.4%
0 5 mm	>3%	>5%	>8%
0.25 mm	>5%	>8%	>11%

The remedial level corresponds with the minimum standard in the standards for the NHSBSP (NHSBSP, 2005). The target value is a term used by the NHSBSP and represents the approximate performance achievable by one third of units. The contrast steps in some test objects (e.g. TOR(MAX)), are relatively large and so very precise measurements may not be possible and due allowance should be made for this when interpreting measurements in comparison with the rounded numbers in the standards.

8.7 Frequency of testing

Image quality measurements must be made to establish a baseline on commissioning new equipment and whenever there are major changes in the system, e.g. a change in screen–film design. These tests should at least involve a comparison with the appropriate national standards (e.g. using a TOR(MAX) test object). However it is strongly recommended to check image quality using a second and possibly more sensitive test object such as the TOR(MAM). These tests should then be repeated at six monthly intervals.

Routine testing of image quality may also be a part of the routine quality control procedures by the operators of x-ray sets. This is a requirement in the NHSBSP on a weekly basis (NHSBSP, 1999b). The test films may be read locally by radiography staff and/or may be submitted for review as part of regional quality control systems. The film should be compared with any quantitative information and with previous films and data. Any deterioration in image quality will necessitate further investigation.

8.8 Description of test objects

(a) TOR(MAX)

The Department of Medical Physics at Leeds University developed the TOR(MAX) test object in 1988 and this was initially recommended for use in the breast screening programme in the United Kingdom (Cowen and Coleman, 1990). (The test object is currently available from Leeds Test Objects Ltd.) This test object comprises a thin D-shaped Perspex plate (to which additional scatter plates can be added) and incorporates five different types of test detail as shown in Figure 8.7. It is used on top of a stack of Perspex plates, usually 4 cm thick. Two line pair test gratings placed at right angles are used to measure limiting high contrast spatial resolution, and simulated calcifications of different size and at different levels of contrast are provided to assess small detail visibility. An additional bar pattern together with three different sizes of circular detail provides measures of low contrast spatial resolution and low contrast detail detectability. The nominal contrasts of the details of the original test objects (old) are shown in Table 8.1. More recently a new manufacturer (Leeds Test Objects Ltd) has produced a revised version with slightly different nominal contrasts as shown in the 'new' column in Table 8.1. It is therefore important to check the manufacturer's specifications when using these test objects. A stepwedge provides the opportunity to obtain a characteristic curve. TOR(MAX) is designed to provide a quantitative assessment of image quality and is basically a category (iii) test object. The TOR(MAS) test object, an earlier version of this test object, has only one high contrast line pair grating. Experience with TOR(MAX) suggests that it can demonstrate poor performance but may not be sufficiently sensitive to reflect the relatively small variation in performance seen amongst the better x-ray systems.

(b) TOR(MAM)

The TOR(MAM) test object was developed to complement the existing test objects TOR(MAX) and TOR(MAS) (Cowen *et al.*, 1992). Although outwardly similar to the other Leeds test objects, TOR(MAM) includes quite different test detail as shown in Figure 8.8. One half of the test object constitutes a type (ii) test object whereby visible details are counted to determine an image quality 'score'. The details contained in this part of the test object comprise filaments, particles and low contrast discs. The other half simulates the appearance of breast tissue with calcification clusters and may be found useful by clinical staff in a non-quantitative way or to rank films in merit order.

Reports from users suggest that rather more care may be needed in scoring the detail half of this test object than with some others, but that it is sensitive to changes in operating parameters such as tube voltage, focal spot size and screen–film system (Law and Faulkner, 1993; Young and Ramsdale, 1993). Results are also sensitive to processing conditions and to background film density. It may have useful potential for development of scoring systems, and even the best of current mammographic systems do not show all the details it contains.

Figure 8.7. Main details of Leeds TOR(MAX) mammography test object. Reproduced with kind permission of Leeds Test Objects Limited

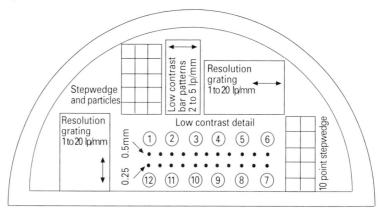

Table 8.1. Nominal contrast of details in old and new TOR(MAX) test objects

	Nominal contrast (%) for 28 kV Mo/Mo*			
	6 mm detail		0.5 mm and 0.25 mm details	
Disc	Old	New	Old	New
1	8.5	8.3	37	41
2	5.5	5.6	33	30
3	3.8	3.9	24	21
4	2.6	2.8	20	16
5	2	2.0	16	11
6	1.6	1.4	11	8.3
7	1.2	1.0	8.3	5.6
8	0.83	0.7	5.4	3.9
9	0.58	0.5	2.7	2.8
10	0.46	0.35	2	2.0
11	0.32	0.25	1.3	1.4
12	0.24	0.15	–	–

*Nominal contrast data provided by Leeds test Objects Ltd and does not include the effect of scatter.

(c) CDMAM

The CDMAM test object, shown in Figure 8.9, is a contrast detail test object developed by the University Medical Centre, Nijmegen. The test object comprises an aluminium base with gold discs of various thickness and diameter attached to a Perspex cover. The test object is used with four Perspex plates, each with a thickness of 10 mm. The discs are arranged in a matrix of 16 rows and 16 columns. The diameters range from 0.06 to 2.00 mm, and the gold thickness ranges from 0.03 to 2.00 μm. Each square contains two identical discs, one in the centre and one in a randomly chosen corner. The test object is designed so an experienced observer with state of the art mammography equipment – screen–film or digital mammography, will detect about half of the discs at standard exposure conditions. The radiation contrast of each thickness of gold is provided in the manual for standard exposure conditions. This permits the full characterisation of threshold contrast across a clinically relevant range of detail sizes.

Image quality and test objects 111

Figure 8.8. Main details of Leeds TOR(MAM) mammography test object. Reproduced with kind permission of Leeds Test Objects Limited

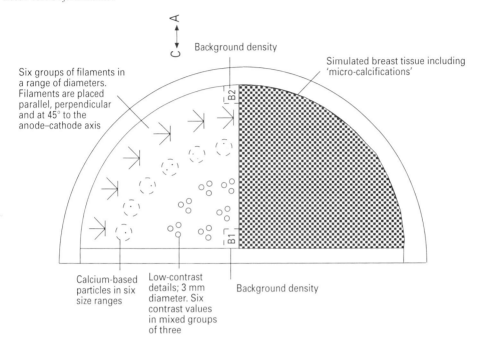

Figure 8.9. Radiograph of CDMAM test object

The test object is, however, time consuming to use and has therefore mainly been used in a research context. Automatic reading of test images created with digital mammography systems may become available to overcome this problem in the future.

(d) Accreditation test object

In the USA a test object is used for accreditation procedures approved by the US Food and Drug Administration. The test object is based on a design originally developed by RMI for accreditation by the American College of Radiology. It consists of an approximately 100 mm square Perspex base block and a cavity filled with a 7 mm thick wax block containing test details comprising simulated tumour masses, nylon fibres and sets of aluminium oxide specks (simulating calcifications). The test object is scored by adding up the number of details detected. The addition of a 4 mm thick Perspex disc on top of the test object is used to assess contrast by measuring the difference in the optical density of the disc compared with the background.

(e) Small field test objects

Small area test objects are desirable for testing image quality on analogue magnification and small field digital systems. Currently there are few test objects of an appropriate size. Where small test objects are not available it is acceptable to image the appropriate details in the TOR(MAX) and TOR(MAM) test objects. For analogue magnification systems this may be possible with a single exposure. However for small field digital systems several images are required to ensure all details have been imaged.

(f) Test patterns for digital displays and printers

As digital devices are becoming increasingly common, so will the use of softcopy reporting and laser film printers. Test images are available to test both the workstations and the laser printers, which are used for reporting. Those workstations that are not used for reporting may not be as critical. Many workstations and laser printers will be purchased with one or more test images that can be used to monitor various aspects of display quality. A common and simple test pattern comprises areas with a range of brightness levels and can be used to monitor changes in the brightness and greyscale characteristics of the monitor or the corresponding optical densities of a laser printed film. The testing of such patterns requires a strict protocol as manual adjustment of the display contrast and brightness will change the measured output. For laser printers, it is often possible to print the equivalent of a sensitometry strip that can be used to feed data back into the system as a means of calibration.

The SMPTE (Society of Motion Picture and Television Engineers) has a common and useful test pattern. This pattern may be included with the workstation or laser imager. It includes a range of greyscale areas from 0 to 100% in 10% steps and both a 0% with 5% inner area and a 100% with 95% inner square (Figure 8.10). Both of the inner squares should be visible. If not, adjustment of the system is required. Spatial resolution and aliasing can also be assessed using the SMPTE test pattern via scoring of a number of black and white bar patterns that are located at the centre and at each corner of the

Image quality and test objects 113

pattern. The bars are both horizontal and vertical and are sized ranging from 2 pixels per bar to 6 pixels per bar. Each bar should be easily distinguishable.

If the digital mammography system is being used as part of a Picture Archive and Communications System (PACS), it is important that the image quality is maintained on each of the networked reporting workstations and laser film printers as there is potential for degradation in image quality through the transmission process. This can be assessed by transmitting the same test pattern, preferably including a resolution pattern, or even a reference image quality image, to each of the workstations and laser printers. As well as ensuring the maintenance of technical image quality, one should also consider the actual transfer of the greyscale pattern. Experience suggests that different laser printers or workstations in the same department can give significantly different greyscale rendition.

Figure 8.10. The SMPTE medical diagnostic imaging test pattern (taken from www.smpte.org/testmat/medical.html)

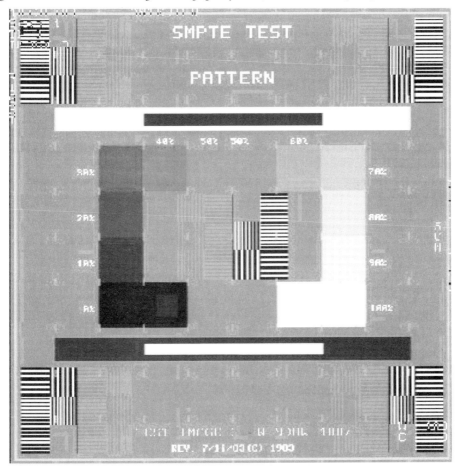

CHAPTER 9

Testing of stereotactic biopsy attachments

9.1 Introduction

There are a number of auxiliary devices used in connection with mammography that may require testing by medical physics staff on commissioning and at regular intervals to ensure adequate operation and radiation safety. The most important of these are stereotactic biopsy attachments. These were first developed for mammography in 1976 and, with the advent of the Breast Screening Programme, these devices, either analogue or digital, are now a standard attachment to most assessment units. Such devices can be used for precise positioning for core biopsies, fine needle aspirations (FNA) and the insertion of guide wires prior to surgical procedures where the lesion is non-palpable. Specially designed dedicated prone biopsy units are also available. Ultrasound is often used with biopsy devices if the lesion is visible on ultrasonic images.

9.2 Basic principles

All currently available stereotactic devices use the same principle. A pair of stereoscopic exposures, either analogue or digital, is made by angling the tube (e.g. 15 or 20°) on either side of the vertical, whilst the woman remains in a fixed position with the breast immobilised by means of a compression paddle. This paddle has a rectangular aperture in its upper surface through which the needle or guide wire can be subsequently introduced vertically downwards. A lateral arm can be attached to produce a horizontal needle guide. An example of a stereotactic device is shown in Figure 9.1. The position of an identified lesion within the breast can be calculated from the stereotactic pair of images.

9.3 Testing of stereotactic localisation and biopsy devices

The x-ray unit and imaging system should be tested as described in Chapter 5. In particular the mechanical safety features given in Section 5.3.2 are important. The following should also be checked:

(i)   The interlocks relating to the operation of the unit in stereotactic mode.

(ii) The fit of any collimators required to be changed manually.

(iii) The cassette fit and the free movement of the cassette from one position to the next.

Testing of stereotactic biopsy attachments 115

Figure 9.1. Illustration of the Siemens mammography stereotactic biopsy unit with a Perspex test object

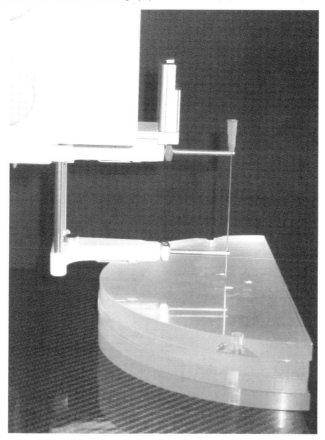

However this chapter concentrates on the need to determine the accuracy of the localisation. The tests and test objects described are appropriate for both analogue and digital systems. Further information on the requirements of stereotactic devices is given by the Department of Health (DH, 1989d).

9.4 Test objects

In view of the need for accuracy of localisation, it is essential for some sort of test object to be available, both to test the accuracy achievable and to allow inexperienced operators to practice and gain confidence in the technique. Several types are available and they can be broadly split into two categories: solid and compressible. Ideally, the test object should be able to be used with both the vertical and horizontal needle guide. Core biopsies are now carried out frequently and are replacing fine needle biopsies. However it is standard practice to routinely test the accuracy of localisation using a needle for fine needle aspirations.

(i) Solid test objects

The principle of a solid test object is to provide a fixed point in space that is visible by eye and on a radiographic image and can be reached by the tip of a needle in the stereotactic device. It can be as simple as a lead shot or thin slice of solder taped to a block of Perspex. More complex designs provide multiple targets at different X, Y, and Z locations on the same exposure.

Test objects of this type are supplied by some manufacturers for checking the operation and accuracy of their localisation device. The older types are pre-drilled blocks of Perspex with a number of holes of depth varying from 5 to 40 mm, and which have a high contrast object at the base of each hole (e.g. a small slice of solder or sub-millimetre lead shot). The disadvantage of this is that the drilled hole tends to guide the needle into position. An 'open' design, like the stepped block of Perspex shown in Figure 9.2 or the one provided with Siemens Opdima, is preferable to the drilled block of Perspex. These designs allow for different needle bores and also do not constrain the needle to approach vertically.

Figure 9.2. An example of a solid Perspex test object

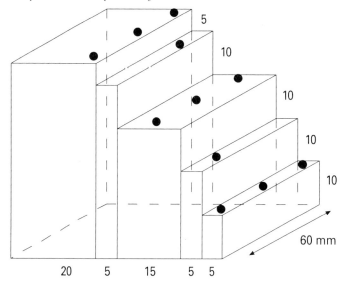

One suggested stereotactic localisation test object (Burch, 2000) consists of a stepped block of Perspex with small pieces of solder fixed to it to represent breast lesions (Figure 9.2). The pieces of solder are approximately 1 mm in diameter and as thin as possible. They are positioned so as to enable measurements of positional accuracy to be made across the whole field of view. It is helpful if each target is slightly different or identifiable, so that one can be sure of selecting the same target in each image of the stereo pair.

These types of test objects are useful for checking the constancy of the positioning process but do not accurately simulate breast tissue. An ideal test object would be compressible and allow the needle to flex.

(ii) Compressible test objects

A variety of different materials may be used which more accurately simulate breast tissue by being compressible and allowing the needle to flex. Fruit such as grapefruit have been used for a number of years but are clearly non-reproducible, messy to use, and have a limited life. Other materials include rubber or plastic orthopaedic foam used either in a single breast-sized piece, or built up in 5–10 mm thick layers (e.g. adhesive orthopaedic foam), which may have sub-millimetre lead shot introduced between the layers. Such test objects have been described by Brennan (1993) as follows.

(a)   Orthopaedic foam

This test object consists of a block of orthopaedic foam material (approximately 100×100×50 mm) containing sub-millimetre lead shot. It can be made from a self adhesive plastic foam sheet (e.g. 'Plasterzote') obtainable from any orthopaedic department or plaster room, using eight sheets of 6 mm thick Plasterzote stuck together with occasional pieces of lead shot sandwiched in each layer. Consequently, the small lead shot is distributed randomly in different (X, Y, Z) positions in the block; individual pieces can be localised using the standard procedure, provided they can be distinguished from each other. Irregular pieces of solder may be preferable.

This test object offers some resistance to the needle and hence is superior to the solid Perspex test object in this respect, and has many more localisation sites. A disadvantage is that the needle puncture sites soon become worn with repeated use, although it can be inverted to extend its use. Further, the lead shot can cause the needle to flex if contact is made.

(b)   'Soft ball' test object

Soft plastic or rubber foam (e.g. a child's soft ball) of approximately 70 mm diameter may be impregnated with sub-millimetre lead shot. Since this may be rotated in any direction the whole surface area may be used for puncture sites, making it less liable to become worn with age. The lead shot may be inserted using a 20 gauge (0.9 mm) lumbar puncture needle. This has an internal diameter of approximately 600 µm, so smaller diameter lead shot must be used.

(iii) Biopsiable test objects

Some manufacturers provide test objects that consist of a gel filled pliable sack containing simulated tumours, which are radio-opaque and contain a liquid dye which can be aspirated by a needle (Jacobsen, 1993). These test objects are most useful for operator training, rather than routine verification of the accuracy of the system as both the number of 'tumours' and needle punctures are limited.

9.5 Measurement procedure

The clinical users of this equipment should be testing it on a regular basis, so it may be appropriate for the physics services to review this data rather than check the accuracy routinely. However, if the unit is only rarely used, physics services should be testing the unit on at least an annual basis.

Following the manufacturer's instructions, attach the stereotactic device to the mammography unit, including the film holder and diaphragms that may be required. Place the test object on the breast platform and bring the appropriate compression paddle into contact with its upper surface (applying the normal amount of compression with 'soft' test objects). Care is needed with placement of the test object to ensure that the holes or lead shot are in the area included in the film.

The manufacturer's instructions should be followed in detail. The procedure below highlights some of the important points. The loaded cassette is first inserted in the special cassette holder so that it is in one of its extreme positions. If a digital detector is selected this is fixed centrally in the beam. The tube is angled to its corresponding extreme position (usually approximately 15° off the vertical) and an exposure made using the usual AEC settings for stereotaxis. With thin foam test objects it may be better to give a small manual exposure, e.g. 5 mAs. The tube is then angled to the other extreme position, the cassette moved to the other side of the holder, and a second exposure made. For GE equipment, tube and cassette are on the same side during each exposure; with Siemens and Lorad equipment they are on opposite sides. In an analogue system, the film is processed in the normal way and the processed film placed on the viewing box part of the equipment, aligning the film with markers where appropriate and clipping into position. A digital detector will display both images on the monitor, and possibly a scout view also. For both types of imaging devices, the system is zeroed by positioning cursors on the pair of reference marks on the film.

To localise the target, move the cursors to a pair of 'speck' images. On the stereo views the correct lead shot is usually easy to determine as the stereo images are always in the same Y-plane. Without the needle holders in position, select the needle length closest to the needle to be used. Needles are often not a precise standard size, e.g. 87 mm instead of nominal 90 mm, in which case all Z values will then read 3 mm high.

X, Y and Z co-ordinates are then displayed on the side of the x-ray set or the viewing box. Set each co-ordinate on the x-ray set to zero, and insert a needle of the correct length into the needle carriage. It is important to check that the correct needle guide size has been used as otherwise the needle will not fit tightly and some movement will be allowed. The tip of the needle should just touch the selected marker and at the same time the needle carriage should just touch the needle handle. If not, adjust the X and Y co-ordinates and note the deviation from zero. If the needle carriage does not touch the needle handle (i.e. the handle sits clear of the carriage) then raise the carriage until they just touch and record the approximate error in the Z co-ordinate.

If a foam test object is used, then the position of the tip of the needle relative to the lead marker cannot be seen and confirmation images will need to be taken. Localisation should be performed for a range of sites from the most superficial to the deepest. A straight needle from current stock should be used, and rotated whilst inserting through the puncture site to the stop position to avoid flexion. On each confirmation image the distance d from the needle tip to the lead shot centre is measured using a ×10 magnifier with a graticule and the maximum value recorded. The (X, Y, Z) co-ordinates are also recorded and from the Z co-ordinate the depth can be determined. Repeat the two images to confirm the accuracy of the position of the needle tip. The procedure should be repeated for a range of sites, including both superficial and deep.

Equipment: Perspex test object, or pliable test object containing lead shot at different depths

Remedial: error in alignment of more than 1 mm in X and Y or more than 3 mm in Z; the manufacturers may claim ±0.5 mm, which may be slightly tighter than is clinically required

Suspension: professional judgement in discussion with the radiologist/user

9.6 Problems

A number of difficulties may arise, especially with the solid Perspex block test object. Despite these problems, this method may still give very useful results. The main points to note are as follows:

(i) The remedial level may be difficult to apply. Action might depend on the view of the radiologist, with whom any errors should be discussed. He/she may require tighter tolerance on X and Y than on Z; not only does Z depend on the bevelled needle tip, but tumours often move down as the needle is inserted, so that a small systematic 'error' in the right direction may actually be useful, and may be built into the system.

(ii) The Z value displayed may not be zero when the needle tip is at the lead shot. Apart from the problem of needle length described already, this may be because the needle tip is bevelled (Figure 9.3). In the clinical situation the needle will tend to draw cells from about 1–2 mm above its tip, so the needle tip position may have been deliberately set to take this into account.

Figure 9.3. Bevelled tip of needle (2–3 mm)

(iii) For Siemens units, errors will be seen if the two needle carriages are not aligned vertically. This must be checked at the time of measurement.

(iv) If a solid drilled Perspex block is used, the needle may not line up with the hole when X and Y are both on zero. Adjust X and Y positions to bring needle in line with the hole – do this by eye until the needle drops into the hole. Read and record X and Y values displayed.

(v) Again if a solid drilled Perspex block is used, the needle may fall roughly into the hole, but not fall entirely freely throughout its length when the needle carriage is lowered. This may mean a bent needle, a non-vertical hole, or a non-vertical carriage. The last named is a problem to which Siemens units are prone, since their carriages consist of a pair of horizontal carrier bars with holes which must be vertically aligned. One solution is to check X and Y as above, and then offset one of these and check Z with the needle tip just resting on the block and the carriage just supporting the needle handle (Figure 9.4). The Z reading should then match the depth of the hole (h).

Figure 9.4. Checking depth when there are misalignment problems. Needle tip resting on upper surface of block when Z=h

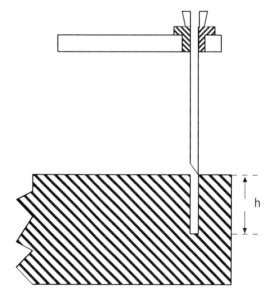

CHAPTER 10

Testing of specimen cabinets used in mammography

10.1 Introduction

In breast imaging, specimen x-ray cabinets are used for imaging sections of breast tissue or biopsy samples once they have been removed from the patient. They are used to confirm that the sample contains the area of tissue or calcification necessary for diagnosis. Specimen cabinets are commonly found in breast units, in pathology units or close to the operating theatres. Most cabinets are tabletop sized and provide a completely enclosed volume in which the x-rays are generated and exposures made. Specimen cabinets are characteristically used at lower tube voltages and tube currents than mammography units with smaller focal spots made from tungsten. Due to these features, testing the cabinets can be more difficult than testing a mammography unit. The clinical use of the cabinet must be taken into account when performing test procedures, with the aim of producing a good clinical image, high in contrast.

Some specimen cabinets may have an automatic exposure control. This is not always desirable for clinical use in mammography due to the thin sections of tissue being imaged, and the corresponding short exposure time. More repeat exposures may be necessary due to the inconsistency of termination times, than would be needed by using incorrect exposure factors. As patient dose is not relevant in specimen radiography, the only disadvantages with repeating exposures are the cost of the film, processing and staff time.

Routine testing by Medical Physics staff is necessary to ensure adequate operation and radiation safety. As the specifications for specimen cabinets vary considerably between the types and ages of the machines, it is important to read the manufacturer's literature before performing any tests. Attention should be paid to the tube rating and suggested cooling intervals necessary between exposures. Some cabinets require a certain warming up time or number of exposures before clinical use. Tests should be made following installation of the cabinet and annually thereafter. Physicists must also check the equipment when the x-ray tube within the unit is replaced or major dismantling/ servicing is carried out. Records of these tests should be maintained to enable comparisons within and between units. Reference should be made to DH (1991).

10.2 Calibration of equipment

The range of tube voltages encountered in these units may be very wide, e.g. 10–110 kV, accompanied by filtration as low as 0.6 mm Be. Ideally, dosimeters and kV meters used for these tests should be suitably calibrated over the energy range normally used. However, calibration is not routinely available for either the energy range or the filtration level encountered in these units, so measurements should be interpreted with care.

10.3 Safety checks

10.3.1 Safety checks (HSE 2000)

Ensure that:

(i) The unit has a satisfactory key-operated power switch.

(ii) There is a functional 'Mains-on' warning light on the control panel. Note that this generally fulfils the requirement of having a device fitted that warns that one further action will generate x-rays (IPEM, 2002).

(iii) There is an indication of x-rays being emitted that is operational, e.g. a warning light.

(iv) X-ray exposures are *not* possible with the cabinet door just ajar, that exposures terminate when the door is opened, and that they do not recommence when the door is closed unless the 'Start' button is also depressed.

10.3.2 Radiation levels outside the cabinet

Specimen cabinets are often used in areas where there is no expectation of radiation exposure, and in close proximity to other work. It is important therefore, that the radiation level around the cabinet is low. If the dose rates outside the cabinet do not exceed 2.5 µSv/h it should not be necessary to designate either a controlled or supervised area. In practice it is rare to find any significant dose rate, because of the low tube voltage employed.

The dose rate should be measured on all sides of the cabinet during x-ray exposures using maximum kilovoltage settings used clinically. Caution should be applied when testing cabinets that are only ever used at mammographic voltages as using the maximum setting may damage the x-ray tube. Given the long exposure times that can be set, it is often practicable to use a protection level dose rate meter, provided that it is suitable for the low energy x-rays involved.

10.4 Generator performance checks

As exposures are made in a completely enclosed cabinet, devices used to measure the x-ray output or energy need to be small enough to be inserted into the cabinet and battery powered. Specimen mammography clinically uses 15–20 kV, but most cabinets allow 10–50 kV and some 10–110 kV. The tests described below may be desirable, but not necessarily practicable in the clinical situation.

10.4.1 Tube voltage

Using electronic potential dividers to test specimen cabinets is problematic, as the devices are not calibrated at the tube voltages used clinically. Hardening the beam with additional sheets of aluminium placed over the divider can overcome this problem. However, the low dose rate arising from the low tube currents (fixed at one value varying from 3 mA for oldest cabinets to 0.3 mA for the newest cabinets) and artificially increased filtration can produce problems with triggering a divider. If the device is placed nearer the exit port to increase the dose rate, beam coverage may be a problem. It may be more appropriate to use a penetrameter type technique, providing that processing artefacts are not a problem. If measurements are possible, they should be performed at a range of clinically used voltages, ensuring that the voltage measured does not vary with exposure time.

10.4.2 Tube output, linearity and consistency

Many specimen cabinets have no method of delineating the centre of the x-ray beam, hence using narrow beam geometry to measure output is not practical. It may not be possible to position the ion chamber in the centre of the field, or to ensure full coverage by the x-ray beam. The focal spot position is rarely marked in an accessible place, so it is often necessary to determine the maximum focus to image distance from the manual, and use this to determine where the chamber should be positioned. It may be more appropriate to ensure a reproducible position using careful description, rather than an absolute central point. Once the ion chamber has been positioned check the consistency of the output, the linearity with time and the linearity with tube voltage.

10.4.3 Exposure time

With dedicated mammography processors and screened cassettes, clinical irradiation times are typically 1–5 s, although the exposures usually range from 1 to 99 s (integer values only) in order to cope with slower speed imaging systems. The difficulties in measuring exposure times mentioned above also apply here, whether the divider or the ion chamber is used to determine exposure times. It is most important that the exposure is consistent, rather than the absolute irradiation time.

10.4.4 Filtration

Measuring the filtration is probably of little value, except to perform corrections for tube voltage measurements, as there will be no dose to a patient from any of these exposures. Typically there is no filtration other than the beryllium exit window, so an HVL of approximately 0.05 mm Al at 28 kV will be found. Filters should either be suspended from the exit port of the tube or positioned on a jig that will fit into the cabinet with the ion chamber and electrometer.

10.5 Imaging performance checks

10.5.1 Beam size and alignment

The beam size at the specimen must be at least 25 cm in diameter. If there is a light marking the centre of the x-ray beam, a pinhole cylindrical tool can be positioned over the cassette to ensure that the alignment is correct.

10.5.2 Magnification factor

There are usually several different magnifications available in a specimen cabinet, by way of a height adjustable shelf. The maximum magnification factor should be at least ×10. If the magnified images are to be used for tumour size assessments the actual magnifications available should be determined. This can be achieved using a small object such as a coin, placed in each of the positions, imaged onto the same cassette.

10.5.3 Focal spot measurement

Nominal focal spot values range from 0.5 mm for the oldest cabinets to 0.02 mm for the newest cabinets. They can be measured by means of star patterns, slits or pinholes, using the most appropriate spoke angle or hole size for the nominal size. It is difficult to measure small focal spot sizes accurately.

10.5.4 Image quality

Most specimens imaged in mammography are thin. For this reason it is appropriate to use an image quality test object with no Perspex. The test object chosen should, if possible, contain some simulated breast tissue. As with any image quality assessment, a nominated cassette should be used for all assessments. Exposures should be made at the tube voltage used clinically, and it may also be of interest to produce an image at 28 kV to compare with mammography x-ray units. It may be difficult to determine exposure time necessary to produce a film of correct density (1.5–1.9 OD), and a compromise may have to be reached due to the integer time values available. If a digital receptor is available, in addition to film, image quality tests should be performed on both imaging systems.

10.6 Automatic exposure control tests (where fitted)

Select a clinically used tube voltage, and vary the Perspex slab thickness above a cassette (0.5–3 cm) recording exposure times. Ensure Perspex fully covers the AEC detector. Optical densities should ideally be within ±0.2. Assess reproducibility of the AEC by making repeat exposures (say 4 or 5), using the same exposure settings and Perspex thickness. If several tube voltages are used clinically, repeat the exposures at each tube voltage used, with the most appropriate thickness of Perspex.

APPENDIX I

Suggested test frequencies

This Appendix lists all the tests in the order given in the protocol, together with suggested frequencies at which they might be undertaken and an indication of the need to perform these tests at commissioning and following the replacement of the x-ray tube or modification or repair. When drawing up local testing schedules, some consideration of the frequency of use of x-ray units which are not used full-time, and of the frequency of use of auxiliary equipment such as magnification tables, large field tables and digital stereotactic attachments, should be made. Some tests may be undertaken by radiographers or other clinical staff. In some cases, it may be appropriate to just review this data rather than repeating the measurements.

Table AI.1.

Test and section in Chapter 5	Commissioning	New tube	AEC modification	Routine frequency
Electrical safety (5.2)	√			As required
Mechanical safety (5.3)	√		√	6 monthly
Radiation safety (5.4)	√	√	√	6 monthly
Alignment (5.6.3)				
Light to x-ray field	√	√		6 monthly
X-ray field to film	√	√		6 monthly
Film to breast support platform edge	√	√		Annually
Alignment tests and size of field on digital systems	√	√		See NHSBSP report 01/09
Leakage radiation (5.6.4)	√	√		
Compression device (5.6.5)				
Compression force	√			6 monthly
Breast thickness indicator	√			6 monthly
Focal spot dimensions (5.6.6)	√	√		Annually
Tube voltage (5.6.7)	√	√		6 monthly
HVL and filtration (5.6.8)				
All target/filter combinations at 28 kV, comp plate out (legal compliance)	√	√		
Clinical target/filter/kV combinations, comp plate in (dose calculation)	√	√		Annually
Tube output (5.6.9)				
Repeatability, specific output, variation with mAs, comp plate out	√	√		
Variation with kV at clinically used target/filter combinations, comp plate in for dose calculations, all foci	√	√		6 monthly
Grid (system) factor (5.6.10)	√			
Image uniformity (5.7.2)	√	√		Annually
Image uniformity for digital systems	√	√		See NHSBSP report 01/09
AEC system (5.7.3)				
Target density	√	√	√	6 monthly
Repeatability	√	√	√	6 monthly
Constancy with thickness	√	√	√	6 monthly
Constancy with tube voltage	√		As necessary	As necessary
Constancy with other parameters	√		As necessary	As necessary
Density control	√			Annually
Guard timer	√		√	Annually
Exposure time	√	√	√	Annually
AEC tests for digital system (5.7.4)	√	√	√	See NHSBSP report 01/09

Table AI.2.

Test and section in Chapter 6	Commissioning	New film/chemistry	Routine frequency
Film processor sensitometry (6.3.1)	√	√	6 monthly
Developer temperature (6.3.2)	√	√	As required
Transport speed (6.3.3)	√	√	As required
Calibration of densitometers (6.4.1)	√		6 monthly
Inter-comparison of sensitometers (6.4.2)	√		Annually
Screen–film contact (6.5.1)			As required
Relative sensitivity of cassette–screen–film batches (6.5.2)			As required
Darkroom check (6.6)	√		Annually
Illuminators/viewing conditions (6.7)	√		Annually
Image artefacts (6.8)			
System (6.8.1)	√	√	6 monthly
Processing (6.8.2)			As required

Table AI.3.

Other tests	Commissioning	Changes to tube or AEC	Changes to processing system	Routine frequency
Mean glandular dose to the standard breast	√	√	If appropriate	6 monthly
Mean glandular dose to real breasts	√		If appropriate	1–3 years
Limiting high contrast spatial resolution (8.6.1)	√	√	√	6 monthly
Threshold contrast (8.6.2)	√	√	√	6 monthly
Sensitive scoring test object (8.7) e.g. TOR(MAM), CDMAM	√	√	√	6 monthly
Computer monitor (8.8f)	√			6 monthly
Laser imager (8.8f)	√			6 monthly
Stereotactic accuracy (9.4)	√			Annually
Specimen cabinet safety (10.3)	√	√		Annually
Specimen cabinet generator (10.4)	√	√		Annually
Specimen cabinet imaging (10.5)	√	√	√	Annually
Specimen cabinet AEC (10.6)	√	√		Annually

APPENDIX II

Useful data

Table AII.1. Manufacturer's data on mammographic x-ray tubes and foci

Manufacturer	Equipment/ model	Tube type/ make	Nominal focus	Target angle (°)	Tube axis inclination (°)	Manufacturer's reference axis	Measurement method
GE/CGR*	Senographe 500T	GS 502-4	0.1/0.3	-9/0	66	7.5/12	IEC 336/1982
GE/CGR*	Senographe 600T	GS 502-4	0.1/0.3	-9/0	66	7.5/12	IEC 336/1982
GE/CGR*	Senix HF	GS 502-4	0.1/0.3	-9/0	66	7.5/12	IEC 336/1982
GE*	DMR	Maxiray70TH-D	Mo – 0.15/0.3				
			Rh – 0.15/0.3	0	30.5/22.5	8/21.5	IEC 336/1993
GE/Picker	Sureview S	Varian Eimac M101	0.4	16	6.5	0	IEC 336/1982
GE/Picker	Sureview SA	Varian Eimac M101	0.1/0.4	16	6	0	IEC 336/1982
Lorad	M3	Eureka Rad86	0.1/0.3	16	4	X	IEC 336/1982
Lorad	M4	Varian M113	0.1/0.3	10/16	6	6	IEC 336/1993
Lorad	M4	Toshiba E7290AX	0.1/0.3	10/16	6	6	X
Philips	MammoDiagnost M	ROM 10-17	0.6	20	0	0	IEC 336/1982
Philips	MammoDiagnost UC	ROM 20	0.1 or 0.15/0.3	20	0	8	IEC 336/1982
Philips	MammoDiagnost 3000	ROM 21	0.1 or 0.15/0.3	20	0	0	IEC 336/1982
Philips	MammoDiagnost 4000	Varian M113	0.1/0.3	10/16	5.3	0	IEC 336/1993
Philips**	MammoDiagnost	P40MoW100G	Mo – 0.15/0.3				
			W – 0.15/0.3	20	0	0/3	IEC 336/1993
Planmed	Sophie	Toshiba E7236 or E7272	0.1/0.3	16	7	X	X
Planmed	Sophie Classic	Toshiba E7272	0.1/0.3	16	7	X	X
Siemens	Mammomat 2	P49 Mo	0.15/0.3	22	0	0/11	IEC 336/1982
Siemens	Mammomat 2S	P49 Ma CRE	0.4	18	0	9	IEC 336/1982
Siemens	Mammomat C	P49 Mo	0.15/0.3	22	0	0/11	IEC 336/1982
Siemens	Mammomat 3 and 300	P49 Mo G	0.15/0.3	22	0	0/9	IEC 336/1993
Siemens	Mammomat 1000	P40 MoW	Mo – 0.15/0.3				
Siemens	Mammomat 3000	P40 MoW	Mo – 0.15/0.3				
			W – 0.15/0.3	20	0	0/7	IEC 336/1993
Siemens	Mammomat 3000 Nova	P40 MoW	Mo – 0.15/0.3	20	0	0/7	IEC 336/1993
			W – 0.15/0.3	20	0	0/7	IEC 336/1993

X   Details not known, consult supplier/manufacturer.
*   Unique tube design, see manufacturer's drawings.
**   Badged version of Siemens Mammomat 3000 Nova.

Half-value layer/filtration curves

These curves give the relation between HVL and added filtration (molybdenum) for different tube voltages and different thicknesses of added Perspex (0, 2, 3 and 4 mm) to simulate the effect of a compression plate (K Cranley, private communication; see also Cranley, 1991). All show calculated values for an effective target angle of 15° and a constant tube potential.

Figure AII.1. Calculated relationship between HVL and Mo filter thickness for an effective target angle of 15° and constant potential waveform. The assumed filtration is 1 mm Be, 500 mm air and 0 mm Perspex

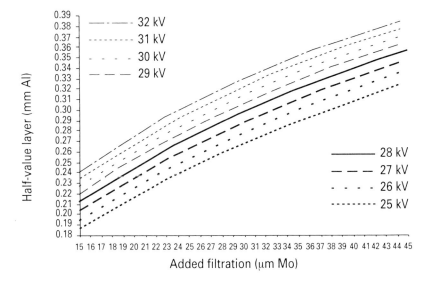

Figure AII.2. Calculated relationship between HVL and Mo filter thickness for an effective target angle of 15° and constant potential waveform of 28 kV. The assumed filtration is 1 mm Be, 500 mm air with 2, 3, or 4 mm Perspex (PMMA)

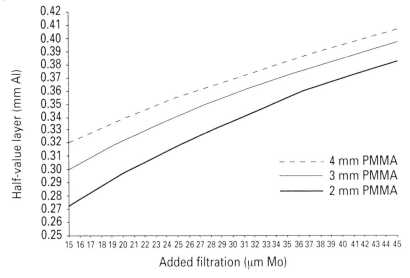

APPENDIX III

Survey of performance measurements

During 2000/01 the National Coordinating Centre for the Physics of Mammography gathered data on the 325 mammography x-ray sets used in the NHS Breast Screening Programme. The data included basic details on the x-ray sets, processors, films, screens and cassettes and quality control measurements made by the local physicists on their most recent survey. This information was provided by physicists throughout the UK. Table AIII.1 is taken from NHS BSP (2003a) and summarises the main performance measurements, which were conducted in accordance with the last edition of this report.

Table AIII.1. Summary of performance measurements on 325 mammography sets (NHSBSP, 2003a)

Parameter	Minimum	25th percentile	Mean	75th percentile	Maximum	Unit
Measured kV at 28 kV set	26.9	27.7	27.9	28.2	29.5	kVp
Output at broad focus at 28 kV Mo/Mo (at 50 cm)	136	182	201	220	282	µGy/mAs
Output per second at FFD at 28 kV Mo/Mo	7.81	11.5	15.5	19.2	29.2	mGy/s
HVL at 28kV with compression paddle	0.30	0.34	0.35	0.365	0.41	mm Al
Broad focus width	0.11	0.30	0.35	0.39	0.63	mm
Broad focus length	0.19	0.42	0.53	0.62	1.09	mm
Fine focus width	0.05	0.10	0.13	0.16	0.45	mm
Fine focus length	0.05	0.12	0.16	0.19	0.37	mm
Separation between film edge and table edge	1	3	3.2	4	6	mm
Overlap of x-ray field at chest wall edge	−2	1	2.0	3	11.2	mm
Maximum compression force (automatically applied)	90	150	170	190	255	N
AEC consistency[a]	0.00%	0.01%	0.75%	1.01%	7.7%	%
AEC error with 2 cm Perspex[b]	−0.35	−0.09	−0.03	0.02	0.22	OD
AEC error with 6 cm Perspex[b]	−0.70	−0.08	−0.02	0.05	0.35	OD
Film density for 4 cm Perspex using AEC at clinical settings	1.31	1.58	1.66	1.74	2.23	OD
Exposure times for 4 cm Perspex	0.16	0.32	0.53	0.64	1.74	s
Exposure times for 6 cm Perspex	0.58	1.21	1.78	2.22	4.12	s
Mean glandular dose to old standard breast (at 28 kV Mo/Mo)	0.65	1.13	1.36	1.61	2.60	mGy
Mean glandular dose to old standard breast (at clinical settings)	0.69	1.20	1.40	1.56	2.60	mGy
High contrast resolution parallel to tube axis[c]	10.0	13.0	14.0	14.5	20.0	lp/mm
High contrast resolution perpendicular to tube axis[c]	11.0	15.0	16.1	17.0	20.0	lp/mm

(a) AEC consistency was measured as the maximum percentage deviation from the mean output for a series of exposures using an ionisation chamber.
(b) AEC error at 2 cm and 6 cm of Perspex calculated as the difference in film density from that measured with 4 cm of Perspex.
(c) High contrast resolution measurements at broad focus and typical clinical setting of 28 kV with grid using Leeds TOR(MAS) or TOR(MAX) on top of 40 mm of Perspex scatter.

APPENDIX IV

Suppliers' addresses

This list may not be complete and should not be interpreted as a recommendation of any particular company or product.

Name and address	Product(s)
Advent Research Materials Ltd Oakfield Industrial Estate Eynsham Oxford OX8 1JA Tel: 01865 884440	Metal foils
Alrad Instruments Ltd Alder House Turnpike Road Industrial Estate Newbury Berkshire RG14 2NS Tel: 01635 46116	Parry densitometers, radiation detectors
Agfa-Gevaert Ltd 27 Great West Road Brentford Middlesex TW8 9AX Tel: 020 8231 4900	Film, screens, cassettes, processors, computed radiography, laser printers
Computerised Imaging Reference Systems (CIRS) Inc. 2428 Almeda Avenue, Suite 212 Norfolk, Virginia 23513 USA Tel: (757) 855 2765	Image quality test objects
Degussa Ltd Winterton House Winterton Way Macclesfield Cheshire SK11 0LP Tel: 01625 503050	Pinholes (focal spot measurement)
Fuji Photo Film (UK) Ltd Unit 12 St Martin's Business Centre St Martin's Way Bedford MK42 0LF Tel: 01234 326780	Film, screens, cassettes, processors, computed radiography

Name and address	Product(s)
Gammex-RMI Ltd Karlsruhe House Queens Bridge Road Nottingham NG1 4FQ Tel: 0115 985 0808	Test equipment, image quality test objects
Goodfellow Ermine Business Park Huntingdon Cambridgeshire PE29 6WR Tel: 01480 424800	Metal foils
Gretag Imaging 2 Newton Court Rankine Road Basingstoke Hampshire RG24 8GF Tel: 01256 370400	Macbeth densitometers
Werner Huttner An der Schwedenschanze 1 Heroldsbach 91336 Germany Tel: (09190) 997442	Resolution (line pair) test objects
IGE Medical Systems Coolidge House 352 Buckingham Avenue Slough Berkshire SL1 4ER Tel: 01753 874000	Mammography x-ray equipment, digital imaging systems, accessories, test equipment
IIE Ingeineurgesellschaft Fuer Industrielle Einwegmedizintechnik Pascal Strasse 7 Aachen 52076 Germany Tel: (02408) 4747	Slit camera (focal spot measurement)
Imaging Equipment Ltd Orpheus House Calleva Park Aldermaston Berkshire RG7 8TA Tel: 0118 981 1828	Test equipment, image quality test objects

Name and address	Product(s)
Integrated Radiological Services Unit 188 Century Building Tower Street Brunswick Business Park Liverpool L3 4BJ Tel: 0151 709 6296	Test equipment, QC software
Kodak Health Imaging Kodak House PO Box 66 Hemel Hempstead Hertfordshire HP1 1JU Tel: 01442 845724	Film, screens, cassettes, processors, laser printers
Konica Plane Tree Crescent Feltham Middlesex TW13 7HD Tel: 020 8751 6121	Film, processors
Leeds Test Objects Ltd Wetherby Road Boroughbridge North Yorkshire YO51 9UY Tel: 01423 321102	Image quality test objects
Medical Imaging Systems Ltd 12 Kingsbury Trading Estate Church Lane Kingsbury London NW9 8AU Tel: 020 8205 9500	Mammography x-ray equipment, digital imaging systems
Philips Medical Systems The Observatory Castlefield Road Reigate Surrey RH2 0FY Tel: 01737 230400	Mammography x-ray equipment

Suppliers' addresses 133

Name and address	Product(s)
Physics Instrument Company Ltd Unit 11 The Old Malthouse Business Park Springfield Road Grantham Lincolnshire NG31 7BG Tel: 01476 569042	Test equipment, image quality test objects
Planmeca Summit House Summit Road Potters Bar Hertfordshire EN6 3EE Tel: 01707 822440	Mammography x-ray equipment
Qados Unit 8 Lakeside Business Park Swan Lane Sandhurst Surrey GU47 9DN Tel: 01252 878999	Specimen radiography cabinets, dosimetry systems
SeeDOS Ltd 26 The Maltings Leighton Buzzard Bedfordshire LU7 4BS Tel: 01525 850670	Test equipment, dosimetry systems
Siemens Medical Engineering Siemens House Oldbury, Bracknell Berkshire RG12 8FZ Tel: 01344 396000	Mammography x-ray equipment, digital imaging systems
Southern Scientific Ltd Scientific House Rectory Farm Road Sompting, Lancing West Sussex, BN15 0DP Tel: 01903 604000	Test equipment, image quality test objects, dosimetry systems

Name and address	Product(s)
Vertec Scientific Ltd 5 Comet House Calleva Park Aldermaston Berkshire RG7 8JA Tel: 0118 981 7431	Dosimetry systems, test equipment
Windense International Crosshill House 30–32 Crosshill Avenue Glasgow G42 8BY Tel: 0141 423 5666	QC software
Wolverson Willenhall Business Park Walsall Street Willenhall West Midlands WV13 2DY Tel: 01902 63733	Mammography x-ray equipment
Xograph Imaging Systems Xograph House Hampton Street Tetbury Gloucestershire GL8 8LD Tel: 01666 501501	Mammography x-ray equipment, digital imaging systems

REFERENCES

Alm Carlsson G, Carlsson C A, Neilsen B and Persliden J 1986 Generalised use of contrast degradation and contrast improvement factors in diagnostic radiology. Application to vanishing contrast *Phys. Med. Biol.* **31** 737–749

Ardran G M, Crooks H E and James V 1969 Testing x-ray cassettes for film-intensifying screen contact *Radiography* **35** 143–145

Barnes G T and Chakraborty D P 1982 Radiographic mottle and patient exposure in mammography *Radiology* **145** 815–821

Barnes G T and Frey G D 1991 *Screen Film Mammography: Imaging Considerations and Medical Physics Responsibilities* (Medical Physics Publishing, Madison, Wisconsin)

Beckett J and Kotre C J 2000 Dosimetric implications of age related glandular changes in screening mammography *Phys. Med. Biol.* **45** 801–813

Beckett J, Kotre C J and Michaelson J S 2003 Analysis of benefit:risk ratio and mortality reduction for the UK Breast Screening Programme *Br. J. Radiol.* **76** 309–320

Beemsterboer P M M, Warmerdam P G, Boer R and de Koning H J 1998 Radiation risk of mammography related to benefit in screening programmes: a favourable balance? *Journal of Medical Screening* **5** 81–87

Bencomo J A, Haus A G, Paulus D D, Hill C E, Logan W W and Johnston D A 1982 Method to study the effect of controlled changes of breast image total system modulation transfer function (MTF) on diagnostic accuracy imaging, in *Application of Optical Instrumentation in Medicine. Proceedings of the Society of Photo-optical Instrumentation Engineers* **347** 84

Brennan A G 1993 A stereotactic test phantom for mammography *Radiat. Prot. Dosim.* **49** 193–195

Brennan A G and Johnson C H 1993 A radiation beam chest wall alignment test tool for mammography *Radiat. Prot. Dosim.* **49** 209–212

Brettle D S, Ward S C, Parkin G J S, Cowen A R and Sumsion H J 1994 A clinical comparison between conventional and digital mammography utilizing computed radiography *Br. J. Radiol.* **67** 464–468

BS 1998 *BS EN 61953:1998 Diagnostic X-ray Imaging Equipment – Characteristics of Mammographic Anti-Scatter Grids* (BSI, London)

BS 2000 *BS EN ISO 9001:2000. Quality Management Systems – Requirements* (BSI, London)

BSI (British Standards Institution) 1981 *Specification for Essential Data to be Supplied with X-ray Tubes and X-ray Tube Assemblies for Medical Use BS6059* (BSI, London)

BSI (British Standards Institution) 1985 *Determination of the Maximum Symmetrical Radiation Field from a Rotating Anode X-ray Tube for Medical Diagnosis BS6628* (BSI, London)

BSI (British Standards Institution) 1989 *Medical Electrical Equipment: General Requirements for Safety. BS5724 Part 1* (BSI, London)

BSI (British Standards Institution) 1991a *Quality Vocabulary: Quality Concepts & related definitions. BS4778 Part 2* (BSI, London)

BSI (British Standards Institution) 1991b *Quality Vocabulary Part 1: Glossary of International Terms BS4778: Part 3, Section 3.2* (BSI, London)

BSI (British Standards Institution) 1994a *Evaluation and Routine Testing in Medical Imaging Departments BS 7725: Section 2.1 Method for Film Processors* (BSI, London)

BSI (British Standards Institution) 1994b *Evaluation and Routine Testing in Medical Imaging Departments BS 7725: Section 2.3 Method for Darkroom Safelight Conditions* (BSI, London)

Bunch P C 1997 *The Effects of Reduced Film Granularity on Mammographic Image Quality* SPIE 3032 302–314

Bunch P C 1999 *Advances in High-speed Mammographic Image Quality* SPIE 3659 120–130

Burch A 2000 Personal communication

Burch A and Goodman D A 1998 A pilot survey of radiation doses received in the United Kingdom Breast Screening Programme *Br. J. Radiol.* **71** 517–527

Burgess A E 1977 Interpretation of star test pattern images *Med. Phys.* **4** 1–8

Campion P J, Burns J E and Williams A 1980 *A Code of Practice for the Detailed Statement of Accuracy* (National Physical Laboratory, Teddington, Middlesex, TW11 0LW)

CEC (Commission of the European Communities) 1989 *Technical and Physical Parameters for Quality Assurance in Medical Diagnostic Radiology: Tolerances, Limiting Values and Appropriate Measuring Methods, BIR Report 18* (British Institute of Radiology, 36 Portland Place, London W1N 4AT)

CEC (Commission of the European Communities) 1996 *European Guidelines on Quality Criteria for Diagnostic Radiographic Images. EUR 16260 EN* (CEC, Brussels)

Cowen A R and Coleman N J 1990 Design of test objects and phantoms for quality control in mammographic screening, in *Physics in Diagnostic Radiology. IPSM Report 61* (IPEM, 230 Tadcaster Road, York YO24 1ES)

Cowen A R, Brettle D S, Coleman N J and Parkin G J S 1992 A preliminary investigation of the imaging performance of photostimulable phosphor computed radiography using a new design of mammographic quality control test object *Br. J. Radiol.* **65** 528

Cowen A R, Launders J H, Jadav M and Brettle D S 1997 Visibility of microcalcifications in computed and screen–film mammography *Phys. Med. Biol.* **42** 1533–1548

Cranley K 1991 Measuring the filtration of mammographic X-ray tubes with molybdenum targets *Br. J. Radiol.* **64** 842–845

Cranley K, Gilmore B J, Fogarty G W A, Desponds L and Sutton D 1997 *Catalogue of Diagnostic X-ray Spectra and Other Data. IPEM Report 78* (IPEM, 230 Tadcaster Road, York YO24 1ES)

Dance D R 1980 The Monte Carlo calculation of integral radiation dose in mammography *Phys. Med. Biol.* **25** 25–37

Dance D R 1988 Diagnostic radiology with X-rays, in *The Physics of Medical Imaging* (ed S Webb) 20–73 (Adam Hilger, Bristol)

Dance D R 1990 Monte Carlo calculation of conversion factors for the estimation of mean glandular breast dose *Phys. Med. Biol.* **35** 1211–1219

Dance D R and Davis R 1983 The physics of mammography, in *Diagnosis of Breast Disease* (ed C A Parsons) 76–100 (Chapman and Hall, London)

Dance D R and Day G 1981 Simulation of mammography by Monte Carlo calculation – the dependence of radiation dose, scatter and noise on photon energy, in *Patient Exposure to Radiation in Medical X-ray Diagnosis* (eds G Drexler, H Eriskat and H Schibilla) EUR7438 227–243 (CEC, Brussels)

Dance D R and Day G J 1984 The computation of scatter in mammography by Monte Carlo methods *Phys. Med. Biol.* **29** 237–247

Dance D R, Persliden J and Alm Carlsson G 1992 Calculation of dose and contrast for two mammographic girds *Phys. Med. Biol.* **37** 235–248

Dance D R, Skinner C L, Young K C, Beckett J R and Kotre C J 2000a Additional factors for the estimation of mean glandular dose using the UK mammography dosimetry protocol *Phys. Med. Biol.* **45** 3225–3240

Dance D R, Thilander Klang A, Sandborg M, Skinner C L, Castellano Smith I A and Alm Carlsson G 2000b Influence of anode/filter material and tube potential on contrast, signal-to-noise ratio and average absorbed dose in mammography: a Monte Carlo study *Br. J. Radiol.* **73** 1056–1067

de Almeida A, Sobol W T and Barnes G T 1999 Characterisation of the reciprocity law failure in three mammography screen–film systems *Med. Phys.* **26** 682–688

Desponds L, Depeursinge C, Grecescu M, Hessler C, Samiri A and Valley J 1991 Influence of anode and filter material on image quality and glandular dose for screen–film mammography *Phys. Med. Biol.* **36** 1165–1182

DH (Department of Health) 1989a *Guidelines on the Establishment of a Quality Assurance Scheme for the Radiological Aspects of Mammography Used for Breast Screening. Report of a Sub-Committee of the Radiological Advisory Committee of the Chief Medical Officer* (Department of Health, PO Box 777, London SE1 6XH)

DH (Department of Health) 1989b *The Harrogate Seminar 15 January 1988. The Report of a Seminar for NHS Officers Engaged in Acceptance Inspections of Radiological Installations* STD/89/03 (Department of Health, PO Box 777, London SE1 6XH)

DH (Department of Health) 1989c *Technical Requirements for the Supply and Installation of Equipment for Diagnostic Imaging and Radiotherapy. Document TRS89* (Department of Health, PO Box 777, London SE1 6XH)

DH (Department of Health) 1989d *Guidance Notes for Health Authorities on Stereotaxic Devices for Fine Needle Aspiration and Positional Marking of Impalpable Breast Lesions. Report STD/89/10* (Department of Health, PO Box 777, London SE1 6XH)

DH (Department of Health) 1991 *Evaluation of Specimen Radiography Cabinets. Report MDD/91/13* (Department of Health, PO Box 777, London SE1 6XH)

DH (Department of Health) 1993 *Guidance Notes for Health Authorities and NHS Trusts on Requirements for Films, Screens and Cassettes Used in Breast Screening Mammography Report MDD/92/43* (Department of Health, PO Box 777, London SE1 6XH)

DH (Department of Health) 1995 *Further Revisions to Guidance Notes for Health Authorities and NHS Trusts on Mammographic X-ray Equipment for Breast Screening. Report MDA/95/40* Medical Devices Agency. Also Addendum 1999 *Provision of rhodium filtration* (Department of Health, PO Box 777, London SE1 6XH)

DH (Department of Health) 2000 *The NHS Cancer Plan: A Plan for Investment. A Plan for Reform* (Department of Health, PO Box 777, London SE1 6XH)

DH (Department of Health) 2001a *Further Revisions to Guidance Notes for Health Authorities and NHS Trusts on Mammographic X-ray Equipment for Breast Screening. Report MDA 01/011, Medical Devices Agency* (Department of Health, PO Box 777, London SE1 6XH)

DH (Department of Health) 2001b *Trex Medical Corporation Lorad Digital Mammography System (DMS), MDA/01/012* (Department of Health, PO Box 777, London SE1 6XH)

DHSS (Department of Health and Social Security) 1985 *Acceptance Inspections of Radiological Apparatus. The Proceedings of a Seminar. Report STD 6A/85/15* (Department of Health, PO Box 777, London SE1 6XH)

DHSS (Department of Health and Social Security) 1986 *Breast Cancer Screening. A Report of a Working Group Chaired by Professor Sir Patrick Forrest* (Department of Health, PO Box 777, London SE1 6XH)

Doi K, Loo L-N and Chan H-P 1982 X-ray tube focal spot sizes: comprehensive studies of their measurement and effect of measured size in angiography *Radiology* **144** 383–393

Evans A, Pinder S, Wilson R, Ellis J (eds) 2002 *Breast Calcification* (Greenwich Medical, London)

Evans D S, Workman A and Payne M 2002 A comparison of the imaging properties of CCD-based devices used for small field digital mammography *Phys. Med. Biol.* **47** 117–135

Evans S H, Bradley D A, Dance D R, Bateman J E and Jones C H 1991 Measurement of small-angle photon scattering for some breast tissues and tissue substitute materials *Phys. Med. Biol.* **36** 7–18

Everson J D and Gray J E 1987 Focal-spot measurements: comparison of slit, pinhole and star resolution pattern techniques *Radiology* **165** 261–264

Fahrig R and Yaffe M J 1994 Optimization of spectral shape in digital mammography: dependence on anode material, breast thickness, and lesion type *Med. Phys.* **21** 1473–1481

Fahrig R, Rowlands J A and Yaffe M J 1995 X-ray imaging with amorphous selenium: detective quantum efficiency of photoconductive receptors for digital mammography *Med. Phys.* **22** 153–160

Faulkner K and Law J 1993 Methods of scoring mammographic phantoms *Radiat. Prot. Dosim.* **49** 183–185

Faulkner K, Law J and Cranley K 1995 Technical note: Perspex blocks for estimation of dose to a standard breast; effect of variation in block thickness *Br. J. Radiol.* **68** 194–196

Feig S A and Ehrlich S M 1990 Estimation of radiation risk from screening mammography: recent trends and comparison with expected benefits *Radiology* **174** 638–647

Freer T W and Ulissey M J 2001 Screening mammography with computer-aided detection: prospective study of 12,860 patients in a community breast center *Radiology* **220** 781–786

Garvican L and Field S 2001 A pilot evaluation of the R2 image checker system and users' response in the detection of interval breast cancers on previous screening films *Clin. Radiol.* **56** 833–837

Geise R A and Palchevsky A 1996 Composition of mammographic phantom materials *Radiology* **198** 342–350

Gingold E L, Wu X and Barnes G T 1995 Contrast and dose with Mo/Mo, Mo/Rh, and Rh/Rh target-filter combinations in mammography *Radiology* **195** 639–644.22

Hammerstein G R, Miller D W, White D R, Masterson M E, Woodward H Q and Laughlin J S 1979 Absorbed radiation dose in mammography *Radiology* **130** 485–491

Haus A G and Jaskulski S M 1997 *The Basics of Film Processing in Medical Imaging.* (Medical Physics Publishing, Madison, Wisconsin, USA)

Heid P 1998 Personal communication

HMSO 1999 *Ionising Radiation Regulations 1999* (Statutory Instrument No. 3232) (Her Majesty's Stationery Office, London)

HMSO 2000 *Ionising Radiation (Medical Exposure) Regulations 2000* (Statutory Instrument No. 1059) (Her Majesty's Stationery Office, London)

HSE (Health and Safety Executive) 1998 *Fitness of Equipment Used for Medical Exposure to Ionising Radiation. Guidance Note, Plant and Machinery 77* (GN PM77)

HSE (Health and Safety Executive) 2000 *Work with Ionising Radiation; Ionising Radiations Regulations 1999 Approved Code of Practice and Guidance* (HSE Books, London)

ICRU (International Commission on Radiation Units and Measurements) 1986 *Modulation Transfer Function of Screen–film Systems. ICRU Report 41* (ICRU, 7910 Woodmont Avenue, Bethesda, MD 20814, USA)

ICRU (International Commission on Radiation Units and Measurements) 1995 *Medical Imaging – The Assessment of Image Quality. ICRU Report 54* (ICRU, 7910 Woodmont Avenue, Bethesda, MD 20814, USA)

ICRP (International Commission on Radiological Protection) 1982 *Protection of the Patient in Diagnostic Radiology. ICRP Publication 34. Annals of the ICRP 9 No. 2/3* (Pergamon Press, Oxford)

ICRP (International Commission on Radiological Protection) 1987 *Statement from the 1987 Como Meeting of the ICRP. Annals of the ICRP 17 No. 4 (ICRP Pub 52)* (Pergamon Press, Oxford)

IEC (International Electrotechnical Commission) 1988 *Quality Assurance in Diagnostic X-ray Departments Part 1: General Aspects* (Draft report)

IEC (International Electrotechnical Commission) 1993a *Determining the Characteristics of Focal Spots in Diagnostic X-ray Tube Assemblies for Medical Use. IEC 336, 1993, BS6530, 1994*

IEC (International Electrotechnical Commission) 1993b *Evaluation and Routine Testing in Medical Imaging Departments – Part 1: General Aspects. IEC/TR2 61223-1 1993*

IPEM (Institute of Physics and Engineering in Medicine) 1996 *Report 32 Part IV, second edition. Measurement of the Performance Characteristics of Diagnostic X-ray Systems used in Medicine – X-ray Intensifying Screens, Films, Processors and Automatic Exposure Control Systems* (IPEM, 230 Tadcaster Road, York YO24 1ES)

IPEM (Institute of Physics and Engineering in Medicine) 1997 *Recommended Standards for the Routine Performance Testing of Diagnostic X-Ray Imaging Systems, first edition. Report No. 77* (IPEM, 230 Tadcaster Road, York YO24 1ES)

IPEM (Institute of Physics and Engineering in Medicine) 1998a *The Critical Examination of X-ray Generating Equipment in Diagnostic Radiology. Report No. 79* (IPEM, 230 Tadcaster Road, York YO24 1ES)

IPEM (Institute of Physics and Engineering in Medicine) 1998b *Quality Control in Magnetic Resonance. Report 80* (IPEM, 230 Tadcaster Road, York YO24 1ES)

IPEM (Institute of Physics and Engineering in Medicine) 2002 *Medical and Dental Guidance Notes* (ed. Allisy-Roberts P J) (IPEM, 230 Tadcaster Road, York YO24 1ES)

IPSM (Institute of Physical Sciences in Medicine) 1989 *The Commissioning and Routine Testing of Mammographic X-ray Systems, first edition. Report No. 59* (IPEM, 230 Tadcaster Road, York YO24 1ES) (out of print)

IPSM (Institute of Physical Sciences in Medicine) 1994 *The Commissioning and Routine Testing of Mammographic X-ray Systems, second edition. Report No. 59* (IPEM, 230 Tadcaster Road, York YO24 1ES) (out of print)

Jacobson D R 1993 A phantom for stereotactic needle biopsy *Radiat. Prot. Dosim.* **49** 197–198

Jacobson D 1994 Practical tools for the assessment of a mammographic focal spot *Med. Phys.* **21** 906

Jarlman O, Samuelsson L and Braw M 1991 Digital luminescence mammography. Early clinical experience *Acta. Radiol.* **32** 110–113

Jennings R J and Fewell T R 1979 Filters – photon energy control and patient exposure, in *Reduced dose mammography* (eds W W Logan and E P Muntz) 212–222 (Masson Publishing, New York, USA)

Johns P C and Yaffe M J 1987 X-ray characterisation of normal and neoplastic breast tissues *Phys. Med. Biol.* **32** 675–695

Keevil S F, Lawinski C P and Morton E J 1987 Measurement of the performance characteristics of anti-scatter grids *Phys. Med. Biol.* **32** 397–401

Kidane G, Speller R D, Royle G J and Hanby A M 1999 X-ray scatter signatures for normal and neoplastic breast tissues *Phys. Med. Biol.* **44** 1791–1802

Kimme-Smith C 1991 Mammography screen–film selection, film exposure and processing, in Barnes and Frey (1991) 135–158

Kimme-Smith C, Bassett L W and Gold R H 1988 Focal spot size measurements with pinhole and slit for microfocus mammography units *Med. Phys.* **15** 298–303

Kimme-Smith C, Bassett L W and Gold R H 1989 A review of test objects for the calibration of resolution, contrast and exposure *Med. Phys.* **16** 758

Klein R, Aichinger H, Dierker J, Jansen J T M, Joite-Barfuß S, Säbel M, Schulz-Wendtland R and Zoetelief J 1997 Determination of average glandular dose with modern mammography units for two large groups of patients *Phys. Med. Biol.* **42** 651–671

Kotre C J, Robson K J and Faulkner K 1993 Assessment of x-ray field alignment in mammography *Br. J. Radiol.* **66** 155–157

Law J 1991 A new phantom for mammography *Br. J. Radiol.* **64** 116–120

Law J 1995 Risk and benefit associated with radiation dose in breast screening programmes *Br. J. Radiol.* **68** 870–876

Law J 1997 Cancers detected and induced in mammographic screening: new screening schedules and younger women with family history *Br. J. Radiol.* **70** 62–69

Law J and Faulkner K 1993 A critical comparison of five different mammographic phantoms under different radiographic conditions *Radiat. Prot. Dosim.* **49** 179–181

Law J and Faulkner K 2001 Cancers detected and induced, and associated risk and benefit, in a breast screening programme *Br. J. Radiol.* **74** 1121–1127

Law J and Faulkner K 2002 Concerning the relationship between benefit and radiation risk, and cancers detected and induced, in a breast screening programme *Br. J. Radiol.* **75** 678–684

Maidment A D A and Yaffe M J 1994 Analysis of the spatial-frequency dependent DQE of optically coupled digital mammography detectors *Med. Phys.* **21** 721–729

MDA (Medical Devices Agency) 1998 *Report MDA/98/10. Konica Radiographic Screens and Films, Part 2: Image Quality Measurements* (Medical Devices Agency, London)

MDA (Medical Devices Agency) 1999a Report MDA/98/52. *Further Revisions to Guidance Notes for Ultrasound Scanners Used in the Examination of the Breast, With Protocol for Quality Testing* (Medical Devices Agency, London)

MDA (Medical Devices Agency) 1999b *Report MDA/98/57. Revised Guidance Notes for Health Authorities on Illuminators for Mammograms for Breast Cancer Screening and Assessment* (Medical Devices Agency, London)

MDA (Medical Devices Agency) 2001 *Report MDA/01/011. Further Revisions to Guidance Notes for Health Authorities and NHS Trusts on Mammographic X-ray Equipment for Breast Screening* (Medical Devices Agency, London)

Meeson S and Young K C 2001 *Image Features of True Positive and False Negative Screening Mammograms. NHSBSP Report 01/07* (NHSBSP Publications, Sheffield)

Meeson S, Young K C, Rust A, Wallis M G, Cooke J and Ramsdale M L 2001 Implications of using high contrast mammography X-ray film screen combinations *Br. J. Radiol.* **74** 825–835

Millis R R, Davis R and Stacey A J 1976 The detection and significance of calcifications in the breast: a radiological and pathological study *Br. J. Radiol.* **49** 12–26

Moores B M 1991 A new phantom for mammography *Br. J Radiol.* **64** 639

Moores B M, Hufton A P, Wrigley C, Asbury D L and Ramsden J A 1979 A quantitative evaluation of film and film/screen combinations for mammographic examination *Br. J. Radiol.* **52** 626–633

Moores B M, Henshaw E T, Watkinson S A and Pearcey B J 1987 *Practical Guide to Quality Assurance in Medical Imaging* (John Wiley, Chichester)

Moy J P 2000 Signal-to-noise ratio and spatial resolution in x-ray electronic imagers: is the MTF a relevant parameter *Med. Phys.* **27** 86–93

NCRP (National Council on Radiation Protection and Measurements) 1986 *Mammography – A User's Guide. Report 85* (NCRP Publications, 7910 Woodmont Avenue, Bethesda, MD 20814, USA)

NHSBSP (National Health Service Breast Screening Programme) 1995 *Guidance Notes on Mammographic X-ray Equipment. Selection; Maintenance; Suspension from Use; Replacement. NHSBSP Publication No. 32* (NHSBSP Publications, Sheffield)

NHSBSP (National Health Service Breast Screening Programme) 1999a *Systematic Management of Quality for Breast Screening Units – A Practical Approach to Quality Management. NHSBSP Publication No. 34 (Part II)* (NHSBSP Publications, Sheffield)

NHSBSP (National Health Service Breast Screening Programme) 1999b *A Radiographic Quality Control Manual for Mammography. Report 21, second edition* (NHSBSP Publications, Sheffield)

NHSBSP (National Health Service Breast Screening Programme) 2000 *Guidelines on Quality Assurance Visits. NHSBSP Publication No. 40* (NHSBSP Publications, Sheffield)

NHSBSP (National Health Service Breast Screening Programme) 2001 *Commissioning and Routine Testing of Small-field Digital Mammography Systems. NHSBSP Report 01/09* (NHSBSP Publications, Sheffield)

NHSBSP (National Health Service Breast Screening Programme) 2003a *Performance of Mammographic Equipment in the UK Breast Screening Programme in 2000/2001 NHSBSP Publication No. 56* (NHSBSP Publications, Sheffield)

NHSBSP (National Health Service Breast Screening Programme) 2003b *Guidance on the Electrical and Mechanical Safety Testing of Mammographic X-ray Equipment. Report 03/01* (NHSBSP Publications, Sheffield)

NHSBSP (National Health Service Breast Screening Programme) (2005, forthcoming) *Consolidated Guidance on Standards for the NHS Breast Screening Programme* (NHSBSP Publications, Sheffield)

Nishikawa R M and Yaffe M J 1985 Signal to noise properties of mammographic screen–film systems *Med. Phys.* **12** 32–39

Ostrum B J, Becker W and Isard H J 1973 Low dose mammography *Radiology* **109** 323–326

Parsons C A (ed) 1983 *Diagnosis of Breast Disease* (Chapman and Hall, London)

Ramsdale M L 1989 Image quality and test phantoms in mammography, in *Optimisation of Image Quality and Patient Exposure in Diagnostic Radiology. BIR Report 20* (ed B M Moores, F E Stieve, H Eriskat, H Schibilla) (British Institute of Radiology, 36 Portland Place, London W1N 4AT)

Ramsdale M L, Hiles P A and Lawinski C P 1989 An alignment test jig for measuring the effective focal spot size of mammographic X-ray tubes *Br. J. Radiol.* **62** 484–486

Rezentes P S, de Almeida A and Barnes G T 1999 Mammography grid performance *Radiology* **228** 227–232

Robinson A and Underwood A C 1991 Scoring of image quality phantom films in mammography *Br. J. Radiol.* **64** 639

Robson K J 2003 Film viewing conditions in x-ray mammography and their effect on observer performance (PhD Thesis, University of Newcastle upon Tyne)

Robson K J, Kotre C J and Faulkner K 1992 The determination of total filtration on mammographic X-ray sets *Br. J. Radiol.* **65** 334–338

Rowlands J A 2002 The physics of computed radiography *Phys. Med. Biol.* **47(23)** R123–R166

Säbel M and Aichinger H 1996 Recent developments in breast imaging *Phys. Med. Biol.* **41** 315–368

Säbel M, Willgeroth F, Aichinger H and Dierken J 1986 X-ray spectra and image quality in mammography *Electromedica* **54** 158–165

Schnall M D 2001 Application of magnetic resonance imaging to early detection of breast cancer *Breast Cancer Research* **3** 17–21

Sexton C, Olson L, Andre M, Olson A, Horton S, Lin S and Sverdrup L 2000 Preliminary clinical evaluation of a digital diagnostic imaging system using a-Se on CMOS *5th International Workshop on Digital Mammography* (Toronto, Canada)

Skubic S E, Yagan R, Oravec D and Shah Z 1990 Value of increasing film processing time to reduce radiation dose during mammography *Am. J. Roent.* **155** 1189–1193

Smilowitz L, Rosen D, Qian H, Phllips W, Stanton M, Stewart A, Mangiafico P, Simoni P and Williams M 1998 A CCD based digital detector for whole-breast digital mammography *4th International Workshop on Digital Mammography* (University of Nijmegen, The Netherlands)

Spiegler P and Breckinridge W C 1972 Imaging of focal spots by means of the star test pattern *Radiology* **102** 679–684

Stevels A L N 1975 New phosphors for x-ray screens *Medicamundi* **20** 12–22

Sutton D G and Williams J R 2000 *Radiation Shielding for Diagnostic X-rays* (British Institute of Radiology, 36 Portland Place, London W1N 4AT)

Tabar L and Haus A G 1989 Processing mammographic films: technical and clinical considerations *Radiology* **173** 65–69

Takahashi K, Kohda K, Miyahara J, Kenemitsu Y, Amitani K and Shionoya S 1984 Mechanism of photostimulated luminescence in $BaFX:Eu^{2+}$ (X=Cl, Br) phosphors *Journal of Luminescence* **31,32** 266–268

Tesic M M, Piccaro M F and Munier B 1999 Full field digital mammography scanner *Eur. J. Radiol.* **31** 2–17

Thijssen M A O, Thijssen H O M, Merx J L, Lindejen J M and Bijkerk K R 1989 A definition of image quality: the image quality figure, in *Optimization of Image Quality and Patient Exposure in Diagnostic Radiology. BIR Report 20* (ed R M Moores, F E Stieve, H Eriskat, H Schibilla) (British Institute of Radiology, 36 Portland Place, London W1N 4AT)

Thijssen M, Veldkamp W, Engen R van, Swinkels M, Karssemeijer, Hendriks J 2001 Comparison of the detectability of small details in a film-screen and a digital mammography system by the imaging of a new CDMAM-phantom, in *Proceedings of IWDM 2000 the 5th International Workshop on Digital Mammography* (ed Yaffe M) (Medical Physics Publishing, Madison Wisconsin, USA)